Machine Learning and the Internet of Medical Things in Healthcare

Machine Learning and the Internet of Medical Things in Healthcare

Edited by

Krishna Kant Singh
*Professor, Faculty of Engineering & Technology,
Jain (Deemed-to-be University), Bengaluru, India*

Mohamed Elhoseny
*College of Computer Information Technology,
American University in the Emirates, Dubai, United Arab Emirates*

Akansha Singh
*Department of CSE, ASET,
Amity University Uttar Pradesh, Noida, India*

Ahmed A. Elngar
*Faculty of Computers and Artificial Intelligence,
Beni-Suef University, Beni Suef City, Egypt*

Academic Press is an imprint of Elsevier
125 London Wall, London EC2Y 5AS, United Kingdom
525 B Street, Suite 1650, San Diego, CA 92101, United States
50 Hampshire Street, 5th Floor, Cambridge, MA 02139, United States
The Boulevard, Langford Lane, Kidlington, Oxford OX5 1GB, United Kingdom

Copyright © 2021 Elsevier Inc. All rights reserved.

No part of this publication may be reproduced or transmitted in any form or by any means, electronic or mechanical, including photocopying, recording, or any information storage and retrieval system, without permission in writing from the publisher. Details on how to seek permission, further information about the Publisher's permissions policies and our arrangements with organizations such as the Copyright Clearance Center and the Copyright Licensing Agency, can be found at our website: www.elsevier.com/permissions.

This book and the individual contributions contained in it are protected under copyright by the Publisher (other than as may be noted herein).

Notices

Knowledge and best practice in this field are constantly changing. As new research and experience broaden our understanding, changes in research methods, professional practices, or medical treatment may become necessary.

Practitioners and researchers must always rely on their own experience and knowledge in evaluating and using any information, methods, compounds, or experiments described herein. In using such information or methods they should be mindful of their own safety and the safety of others, including parties for whom they have a professional responsibility.

To the fullest extent of the law, neither the Publisher nor the authors, contributors, or editors, assume any liability for any injury and/or damage to persons or property as a matter of products liability, negligence or otherwise, or from any use or operation of any methods, products, instructions, or ideas contained in the material herein.

British Library Cataloguing-in-Publication Data
A catalogue record for this book is available from the British Library

Library of Congress Cataloging-in-Publication Data
A catalog record for this book is available from the Library of Congress

ISBN: 978-0-12-821229-5

For Information on all Academic Press publications
visit our website at https://www.elsevier.com/books-and-journals

Publisher: Mara Conner
Acquisitions Editor: Chris Katsaropoulos
Editorial Project Manager: Ruby Smith
Production Project Manager: Swapna Srinivasan
Cover Designer: Miles Hitchen

Typeset by MPS Limited, Chennai, India

Contents

List of contributors xiii

CHAPTER 1 Machine learning architecture and framework...........1

Ashish Tripathi, Arun Kumar Singh, Krishna Kant Singh, Pushpa Choudhary and Prem Chand Vashist

1.1 Introduction 1

    1.1.1 Machine learning classification 2

1.2 Architecture of machine learning 4

    1.2.1 Data acquisition 5

    1.2.2 Data processing 5

    1.2.3 Data modeling 9

    1.2.4 Execution (model evaluation).......................... 12

    1.2.5 Deployment 12

1.3 Machine learning framework 12

    1.3.1 Features of ML framework 12

    1.3.2 Types of ML framework.................................... 15

1.4 Significance of machine learning in the healthcare system 17

    1.4.1 Machine-learning applications in the healthcare system 18

1.5 Conclusion...19

    References... 20

CHAPTER 2 Machine learning in healthcare: review, opportunities and challenges 23

Anand Nayyar, Lata Gadhavi and Noor Zaman

2.1 Introduction 23

    2.1.1 Machine learning in a nutshell 24

    2.1.2 Machine learning techniques and applications................. 25

    2.1.3 Desired features of machine learning.............................. 28

    2.1.4 How machine learning works? 29

    2.1.5 Why machine learning for healthcare?........................... 30

2.2 Analysis of domain 32

    2.2.1 Background and related works 32

    2.2.2 Integration scenarios of ML and Healthcare.................... 33

    2.2.3 Existing machine learning applications for healthcare 33

vi   Contents

    2.3 Perspective of disease diagnosis using machine learning 37

        2.3.1 Future perspective to enhance healthcare system using machine learning 38

    2.4 Conclusions 41

    References...41

CHAPTER 3   **Machine learning for biomedical signal processing**.. **47**

Vandana Patel and Ankit K. Shah

    3.1 Introduction 47

    3.2 Reviews of ECG signal...48

    3.3 Preprocessing of ECG signal using ML based techniques 51

        3.3.1 Least mean square (LMS).. 54

        3.3.2 Normalized least mean square (NLMS)........................ 54

        3.3.3 Delayed error normalized LMS (DENLMS) algorithm 54

        3.3.4 Sign data least mean square (SDLMS) 55

        3.3.5 Log least mean square (LLMS)................................... 57

    3.4 Feature extraction and classification of ECG signal using ML-based techniques 58

        3.4.1 Artificial neural network (ANN) 59

        3.4.2 Fuzzy logic (FL) 60

        3.4.3 Wavelet transforms 61

        3.4.4 Hybrid approach.. 62

    3.5 Discussions and conclusions 63

    References...63

CHAPTER 4   **Artificial itelligence in medicine** **67**

Arun Kumar Singh, Ashish Tripathi, Krishna Kant Singh, Pushpa Choudhary and Prem Chand Vashist

    4.1 Introduction 67

        4.1.1 Disease 67

        4.1.2 Medicine.. 70

        4.1.3 History of AI in medicine.. 73

        4.1.4 Drug discovery process... 75

        4.1.5 Machine-learning algorithms in medicine..................... 77

        4.1.6 Expert systems 82

        4.1.7 Fuzzy expert systems 83

        4.1.8 Artificial neural networks 83

    4.2 Conclusion...84

    References...85

Contents **vii**

CHAPTER 5 Diagnosing of disease using machine learning **89**

Pushpa Singh, Narendra Singh,
Krishna Kant Singh and Akansha Singh

5.1 Introduction 89

5.2 Background and related work...90

  5.2.1 Challenges in conventional healthcare system................ 91

  5.2.2 Machine-learning tools for diagnosis and prediction....... 91

  5.2.3 Python.. 92

  5.2.4 MATLAB 93

5.3 Types of machine-learning algorithm.............................93

5.4 Diagnosis model for disease prediction..........................95

  5.4.1 Data preprocessing... 95

  5.4.2 Training and testing data set.................................. 95

  5.4.3 Classification technique....................................... 96

  5.4.4 Performance metrics 97

5.5 Confusion matrix..97

5.6 Disease diagnosis by various machine-learning algorithms........98

  5.6.1 Support vector machine (SVM).................................. 98

  5.6.2 K-nearest neighbors (KNN)..................................... 100

  5.6.3 Decision tree (DT) 101

  5.6.4 Naive bayes (NB)... 103

5.7 ML algorithm in neurological, cardiovascular,
and cancer disease diagnosis...104

  5.7.1 Neurological disease diagnosis by machine
  learning... 104

  5.7.2 Cardiovascular disease diagnosis by machine
  learning... 105

  5.7.3 Breast cancer diagnosis and prediction:
  a case study 105

  5.7.4 Impact of machine learning in the healthcare
  industry.. 106

5.8 Conclusion and future scope..107

  References.. 107

**CHAPTER 6 A novel approach of telemedicine for
managing fetal condition based on machine
learning technology from IoT-based wearable
medical device**... **113**

Ashu Ashu and Shilpi Sharma

6.1 Introduction 113

6.2 Healthcare and big data...113

viii   Contents

6.3 Big data analytics 115
6.4 Need of IOT in the healthcare industry....................................116
6.5 Healthcare uses machine learning 117
6.6 Need for machine learning...117
6.7 Cardiotocography 118
6.8 Literature review 119
    6.8.1 Research on revolutionary effect of telemedicine and its history... 119
    6.8.2 Role of machine learning in telemedicine/healthcare......... 120
    6.8.3 Role of big data analytics in healthcare 122
    6.8.4 Challenges faced in handling big data in healthcare/telemedicine.. 123
    6.8.5 Research done on tracing the fetal well-being using telemedicine and machine learning algorithms.... 125
6.9 Methodology...126
    6.9.1 Preprocessing and splitting of data................................ 127
6.10 Evaluation..131
6.11 Conclusion and future work..132
    References... 132

CHAPTER 7 **IoT-based healthcare delivery services to promote transparency and patient satisfaction in a corporate hospital** 135
Ritam Dutta, Subhadip Chowdhury and Krishna Kant Singh

7.1 Introduction 135
7.2 Uses of IoT in healthcare..136
7.3 Main problem area of a corporate hospital 137
    7.3.1 Location.. 138
    7.3.2 Hassle on outpatient services.. 138
    7.3.3 Diagnostic services 138
    7.3.4 Inpatient services 138
    7.3.5 Support and utility services 139
    7.3.6 Coordination in medical section...................................... 139
    7.3.7 Medical record keeping 139
    7.3.8 Transparency 139
    7.3.9 Cost leadership model in market.................................... 140
7.4 Implementation of IoT-based healthcare delivery services.......140
    7.4.1 The work of value chain... 140
7.5 Conclusion...150
    References... 150

Contents ix

CHAPTER 8 Examining diabetic subjects on their correlation with TTH and CAD: a statistical approach on exploratory results 153

Subhra Rani Mondal and Subhankar Das

8.1 Introduction 153
  8.1.1 General application procedure 154
  8.1.2 Medicinal imaging 154
  8.1.3 Big data and Internet of Things 156
  8.1.4 Artificial intelligence (AI) and machine learning (ML) 156
  8.1.5 Big data and IoT applications in healthcare 158
  8.1.6 Diabetes and its types 159
  8.1.7 Coronary artery disease (CAD) 161
8.2 Review of literature 163
8.3 Research methodology 164
  8.3.1 Trial setup 164
8.4 Result analysis and discussion 166
  8.4.1 TTH cannot be 169
8.5 Originality in the presented work 170
8.6 Future scope and limitations 171
8.7 Recommendations and considerations 171
8.8 Conclusion 172
  References 173

CHAPTER 9 Cancer prediction and diagnosis hinged on HCML in IOMT environment 179

G. S. Pradeep Ghantasala, Nalli Vinaya Kumari and Rizwan Patan

9.1 Introduction to machine learning (ML) 179
  9.1.1 Some machine learning methods 179
  9.1.2 Machine learning 179
9.2 Introduction to IOT 181
9.3 Application of IOT in healthcare 182
  9.3.1 Redefining healthcare 183
9.4 Machine learning use in health care 185
  9.4.1 Diagnose heart disease 185
  9.4.2 Diabetes prediction 186
  9.4.3 Liver disease prediction 187
  9.4.4 Surgery on robots 187
  9.4.5 Detection and prediction of cancer 188

x Contents

9.4.6 Treatment tailored 188
9.4.7 Discovery of drugs 190
9.4.8 Recorder of intelligent digital wellbeing 190
9.4.9 Radiology machine learning 190
9.4.10 Study and clinical trial 191
9.5 Cancer in healthcare 192
9.5.1 Methods 192
9.5.2 Result 192
9.6 Breast cancer in IoHTML 193
9.6.1 Study of breast cancer using the adaptive voting algorithm 193
9.6.2 Software development life cycle (SDLC) 193
9.6.3 Parts of undertaking duty PDR and PER 194
9.6.4 Info structure 195
9.6.5 Input stage 195
9.6.6 Output design 195
9.6.7 Responsible developers overview 195
9.6.8 Data flow 196
9.6.9 Cancer prediction of data in different views 196
9.6.10 Cancer predication in use case view 196
9.6.11 Cancer predication in activity view 196
9.6.12 Cancer predication in class view 196
9.6.13 Cancer predication in state chart view 198
9.6.14 Symptoms of breast cancer 198
9.6.15 Breast cancer types 198
9.7 Case study in breast cancer 200
9.7.1 History and assessment of patients 201
9.7.2 Recommendations for diagnosis 201
9.7.3 Discourse 202
9.7.4 Outcomes of diagnosis 203
9.8 Breast cancer algorithm 204
9.9 Conclusion 206
References 207

CHAPTER 10 Parameterization techniques for automatic speech recognition system **209**

Gaurav Aggarwal, Sarada Prasad Gochhayat and Latika Singh

10.1 Introduction 209
10.2 Motivation 210
10.3 Speech production 211

Contents **xi**

10.4 Data collection 214
 10.4.1 Recording procedure...................................... 214
 10.4.2 Noise reduction 215
10.5 Speech signal processing 215
 10.5.1 Sampling and quantization 216
 10.5.2 Representation of the signal in time and
 frequency domain.. 217
 10.5.3 Frequency analysis.. 219
 10.5.4 Short time analysis.. 221
 10.5.5 Short-time fourier analysis............................ 222
 10.5.6 Cepstral analysis 222
 10.5.7 Preprocessing: the noise reduction technique 223
 10.5.8 Frame blocking 225
 10.5.9 Windowing.. 226
10.6 Features for speech recognition 226
 10.6.1 Types of speech features................................ 226
10.7 Speech parameterization 228
 10.7.1 Feature extraction.. 228
 10.7.2 Linear predicative coding (LPC).................... 229
 10.7.3 Linear predictive cepstral coefficients (LPCC)............ 231
 10.7.4 Weighted linear predictive cepstral coefficients
 (WLPCC) 233
 10.7.5 Mel-frequency cepstral coefficients 234
 10.7.6 Delta coefficients 239
 10.7.7 Delta–delta coefficients 240
 10.7.8 Power spectrum density 240
10.8 Speech recognition 242
 10.8.1 Types of speech pattern recognition.............. 244
10.9 Speech classification 244
 10.9.1 Artificial neural network (ANN).................... 245
 10.9.2 Support vector machine (SVM)...................... 245
 10.9.3 Linear discriminant analysis (LDA).............. 246
 10.9.4 Random forest 246
10.10 Summary and discussion..246
 References.. 248

**CHAPTER 11 Impact of big data in healthcare system—a quick
 look into electronic health record systems** **251**
 *Vijayalakshmi Saravanan, Ishpreet Aneja, Hong Yang,
 Anju S. Pillai and Akansha Singh*
11.1 A leap into the healthcare domain..251

11.2	The real facts of health record collection	253
11.3	A proposal for the future	254
11.4	Discussions and concluding comments on health record collection	255
11.5	Background of electronic health record systems	256
	11.5.1 The definition of an electronic health record (EHR)	256
	11.5.2 A short history of electronic health records	257
11.6	Review of challenges and study methodologies	257
	11.6.1 Analyzing EHR systems and burnout	257
	11.6.2 Analyzing EHR systems and productivity	258
	11.6.3 Analyzing EHR systems and data accuracy	259
11.7	Conclusion and discussion	260
	References	261

Index 263

List of contributors

Gaurav Aggarwal
School of Computing and Information Technology, Manipal University Jaipur, Jaipur, India

Ishpreet Aneja
Department of Data Science, Rochester Institute of Technology, Rochester, NY, United States

Ashu Ashu
Department of Computer Science, Amity University, Noida, India

Pushpa Choudhary
Department of Information Technology, G. L. Bajaj Institute of Technology and Management, Greater Noida, India

Subhadip Chowdhury
Durgapur Society of Management Science College, KNU, Asansol, India

Subhankar Das
Researcher and Lecturer, Honors Programme, Duy Tan University, Da Nang, Vietnam

Ritam Dutta
Surendra Institute of Engineering and Management, MAKAUT, Kolkata, India

Lata Gadhavi
IT Department, Government Polytechnic Gandhinagar, Gandhinagar, India

G. S. Pradeep Ghantasala
Department of Computer Science and Engineering, Chitkara University Institute of Engineering & Technology, Chandigarh, India

Sarada Prasad Gochhayat
Virginia Modeling, Analysis and Simulation Centre, Simulation and Visualization Engineering, Old Dominion University, Suffolk, VA, United States

Nalli Vinaya Kumari
Department of Computer Science and Engineering, Malla Reddy Institute of Technology and Science, Hyderabad, India

Subhra Rani Mondal
Researcher and Lecturer, Honors Programme, Duy Tan University, Da Nang, Vietnam

Anand Nayyar
Faculty of Information Technology, Graduate School, Duy Tan University, Da Nang, Vietnam

Rizwan Patan
Department of Computer Science and Engineering, Velagapudi Ramakrishna Siddhartha Engineering College, Vijayawada, India

xiv List of contributors

Vandana Patel
Department of Instrumentation and Control Engineering, Lalbhai Dalpatbhai College of Engineering, Ahmedabad, India

Anju S. Pillai
Department of Electrical and Electronics Engineering, Amrita School of Engineering, Amrita Vishwa Vidyapeetham, Coimbatore, India

Vijayalakshmi Saravanan
Faculty in Department of Software Engineering, Rochester Institute of Technology, Rochester, NY, United States

Ankit K. Shah
Department of Instrumentation and Control Engineering, Lalbhai Dalpatbhai College of Engineering, Ahmedabad, India

Shilpi Sharma
Department of Computer Science, Amity University, Noida, India

Akansha Singh
Department of CSE, ASET, Amity University Uttar Pradesh, Noida, India

Arun Kumar Singh
Department of Information Technology, G. L. Bajaj Institute of Technology and Management, Greater Noida, India

Krishna Kant Singh
Faculty of Engineering & Technology, Jain (Deemed-to-be University), Bengaluru, India

Latika Singh
Ansal University, Gurugram, India

Narendra Singh
Department of Management Studies, G. L. Bajaj Institute of Management and Research, Greater Noida, India

Pushpa Singh
Department of Computer Science and Engineering, Delhi Technical Campus, Greater Noida, India

Ashish Tripathi
Department of Information Technology, G. L. Bajaj Institute of Technology and Management, Greater Noida, India

Prem Chand Vashist
Department of Information Technology, G. L. Bajaj Institute of Technology and Management, Greater Noida, India

Hong Yang
Department of Data Science, Rochester Institute of Technology, Rochester, NY, United States

Noor Zaman
School of Computer Science and Engineering SCE, Taylor's University, Subang, Jaya, Malaysia

CHAPTER 1

Machine learning architecture and framework

Ashish Tripathi[1], Arun Kumar Singh[1], Krishna Kant Singh[2], Pushpa Choudhary[1] and Prem Chand Vashist[1]

[1]Department of Information Technology, G. L. Bajaj Institute of Technology and Management, Greater Noida, India
[2]Faculty of Engineering & Technology, Jain (Deemed-to-be University), Bengaluru, India

1.1 Introduction

In 1959, Arthur Samuel proposed the term Machine Learning (ML). He was the master of artificial intelligence and computer gaming. He stated that machine learning gives a learning ability to computers without being explicitly programmed.

In 1997 a relational and mathematical-based definition of machine learning was given by Tom Mitchell in that a computer program uses experience "e" to learn from some task "t." It applies "p" as a performance measure on "t" that automatically improves with "e."

In recent years, machine learning has been appearing as the most significant technology around the globe to solve many real-life problems. So, it is emerging as a very popular technology among the researchers and industry people for problem solving.

Machine learning is a subdomain of artificial intelligence (AI) that helps machines to automatically learn from experience and improve their ability to take decisions to solve any problem without taking any explicit instructions. The focus of ML is to develop and apply computer programs that are able to learn from the problem domain and make better decisions [1].

The learning process in ML starts with observing and analyzing the data through different techniques, such as using examples, experiences, relying on pattern matching in data, etc., that allows machines to take decisions without any assistance from humans or any other intervention [2].

Machine-learning algorithms take a sample data set as an input, which is also known as the training data set, to build and train a mathematical model (ML system). Input data may include text, numerics, audio, multimedia, or visual things and can be taken from various sources such as sensors, applications, devices, networks, and appliances.

Machine Learning and the Internet of Medical Things in Healthcare. DOI: https://doi.org/10.1016/B978-0-12-821229-5.00005-7
Copyright © 2021 Elsevier Inc. All rights reserved.

The mathematical model is used to extract knowledge from the input data through analyzing itself without any explicit programming intervention.

After processing the data, it gives some response as an output. The output may be in the form of an integer value or a floating-point value [3]. Fig. 1.1 shows the learning of a ML system from data with explicit programming support.

Unlike conventional algorithms, ML algorithms are applicable in different applications, such as computer vision, filtering of emails, ecommerce, health care systems, and many more, to provide effective solutions with high accuracy.

1.1.1 Machine learning classification

Learning can be understood as a process that converts experience into expertise. So, the learning should be meaningful in respect to some task. It is a clearly defined process that takes some inputs and produces an output accordingly.

Machine-learning algorithms have three major classifications that depend on the type and nature of data provided to the system, which are as follows:

1.1.1.1 Supervised learning

Supervised learning starts with a data set which contains both input and expected output data with labels. These labels are used for classification and provide a base for learning [4]. This helps in future data processing.

The term supervised learning is considered to be a system which contains pairs of inputs and outputs to provide training to the machine to correlate inputs and outputs based on certain rules [5].

Indeed, supervised learning is to train the machine how the given inputs and outputs can be mapped or related together. The objective of this learning is to

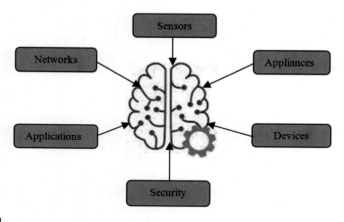

FIGURE 1.1

Learning of a ML system from different data sources.

make the machine capable enough by producing a mapping function to predict the correct output on the given inputs to the system.

Supervised learning is applicable in various applications, such as self-driving cars, chatbots, facial recognition, expert systems, etc.

Artificial Neural Networks [6,7], Logistic Regression [8], Support Vector machine [9], K-Nearest Neighbor [10], and Naïve Bayes Classifier [11] are some examples of supervised learning algorithms.

Unsupervised learning problems are grouped under classification and regression.

- Classification: it is a technique that helps to categorize the inputs into two or more classes based on the features of the inputs. It is used to assign the correct class label to the inputs [12]. In classification, prediction is based on yes or no [13]. A suitable example of classification is spam filtering in which inputs are email messages and the corresponding classes are spam and not spam. Other examples are sentiment analysis to analyze the positive and negative sentiment, labeling of secure and unsecure loans, and to classify whether the person is male or female.
- Regression: in regression, a continuous or real rather than discrete value is obtained as the output [14].

For example, the prediction of house price based on size or quantifying height based on the relative impact of gender, age, and diet.

1.1.1.2 Unsupervised learning

Unsupervised learning is used where the data are not labeled or classified to train the system [15]. There is no labeled sample data available for the training purposes.

This learning system itself works to explore and recognize the patterns and structures in the input data to obtain a predicted value or classification of an object [16].

Unsupervised learning does not claim to give the exact figure of the output.

This form of learning is applied on unlabeled data to train the system to explore and exploit the hidden structure. It helps to categorize the unclassified and unlabeled data based on feature extraction [17]. This learning system works in a real-time scenario and hence the presence of learners is the prime necessity to label and analyze the input data [18].

Thus in this learning, human intervention for analyzing and labeling the unlabeled data is required, which is not found in supervised learning [19].

Unsupervised learning can be understood by an example: an unseen image containing goat and sheep has been given for identifying both separately. Thus in the absence of the information about the features of both the animals, the machine is not able to categorize both. But, we can categorize both the sheep and goat based on their differences, similarities, and patterns. First, we need to separate the pictures of goats and secondly collect all pictures of sheep from the image. Here, in the task, there has not been any training or sample data used to train the machine previously. Mapping of nearest neighbor, value decomposition, self-organizing maps, and k-means clustering are the most used unsupervised learning techniques.

Unsupervised learning problems are grouped into clustering and association problems.

- Clustering: this concept deals with finding a group of uncategorized data based on certain features, patterns, or structures [20]. The clustering algorithms are responsible for identifying groups/clusters of data. For example, considering purchasing behavior of customers to identify a group of customers.
- Association: this allows the establishment of association among data objects from the large set of your data [18,21]. This can be understood by, for example, an association exists if a person that buys an object x then also has a tendency to buy object y.

1.1.1.3 Reinforcement learning

Reinforcement learning involves interaction with the surrounding environment and it applies a trial and error approach to obtain rewards or errors. Algorithms identify those actions which contain the best rewards [22].

Reinforcement learning has three main components: environment, actions, and agent. Here the agent works as a decision-maker, actions denote the steps that are taken by the agent, and the environment is the domain in which the agent does the interaction [23]. To get the maximum level of performance from machine/software agents, this learning helps to decide the ideal behavior automatically for a particular context.

This type of learning technique uses a reinforcement signal that requires feedback for guiding the agent to decide the best among all the available actions. Additionally, it helps to decide to take action, where it is hard to predict the level of severity for certain situations based on goodness or badness [24]. This learning enables machines to learn and play games, drive vehicles, etc.

So, the main objective of reinforcement learning is to apply the best technique to achieve a fast business outcome as early as possible.

1.2 Architecture of machine learning

The required industry interest has been incorporated in the architecture of machine learning.

Thus the objective is to optimize the use of existing resources to get the optimized result using the available data. Also, this helps in predictive analysis and data forecasting in various applications, especially when it is linked with data science technology.

The architecture of machine learning has been defined in various stages. Each stage has different roles and all stages work together to optimize the decision support system.

1.2 Architecture of machine learning

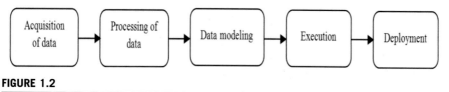

FIGURE 1.2

Phases of machine learning architecture.

The architecture of machine learning has been divided into five stages, such as data acquisition, data processing, data modeling, execution, and deployment. These stages are shown in Fig. 1.2.

The details are as follows:

1.2.1 Data acquisition

It is well-known that the acquisition of data is problem-specific and it is unique for each ML problem. The exact estimation of data is very difficult in order to obtain the optimal utility with respect to machine learning problems. It is very tough to predict what amount of data is required to train the model in the early stage of data acquisition.

In some research activities, it is found that more than two-thirds of the collected data may be useless. Also, at the time of data collection, it is very difficult to understand which portion of the data will be able to provide the significant and correct result before the training begins for a model. Therefore it becomes essential to accumulate and store all kinds of data, whether it is related to structured, unstructured, online, offline, open, and internal. So this phase of data acquisition should be taken very seriously because the success of the ML model training depends on the relevance and quality of data.

Therefore the data acquisition is the first phase in ML architecture that applies to accumulate the essential data from different sources [25], as shown in Fig. 1.3.

The data is further processed by the system to make a decision for solving a given problem. This includes different activities such as gathering comprehensive and relevant data, case-based data segregation, and valid interpretation of data for storing and processing as per the requirements [26].

Actually, the data is gathered from different sources in an unstructured format and every source has a different format which is not suitable for analysis purposes [27].

1.2.2 Data processing

This stage accepts data from the data acquisition layer to apply further processing which includes data integration, normalization, filtering, cleaning, transformation, and encoding of data.

FIGURE 1.3

Data acquisition.

Processing of data also depends on the learning techniques which have been used to solve the problem [28]. For example, in the case of supervised learning, data segregation is performed which creates sample data in several steps. The sample data is further applied to train the system and thus the created sample data is generally called the training data [29].

In unsupervised learning, the unlabeled data is mainly used for analysis purposes. Thus this learning technique mainly deals with unpredictable data that are more complicated and require complex processing as compared to other existing learning techniques [29].

In this case data are grouped into clusters and each cluster belongs to a specific group. Each cluster is formed based on the granularity of the data.

The kind of processing is another factor of data processing which is based on the features and action taken on the continuous data. Also, it may process upon discrete data. Processing on discrete data may need memory bound processing [30].

So, the objective of this stage is to provide a clean and structured data set. Sometimes, this stage is also known as the preprocessing stage. Some major steps come under this phase of ML architecture, which are as follows:

1.2.2.1 Arrangement of data

The stored data are required to be arranged in sorted order with the filtering mechanism. This helps to organize the data in some understandable form.

It becomes very easy to retrieve the required information, which is necessary for the visualization and analysis purposes of the data.

1.2.2.2 Analysis of data

The in-depth analysis starts to understand the data in terms of its type, any missing parts, value of data, correlation among data, and much more, in order to take further action on the data. In data analysis, data is evaluated based on the logical and analytical reasoning to explore and explain each element of the data provided to reach a final conclusion.

1.2.2.3 Preprocessing of data

Data preprocessing is a technique that performs the necessary conversion of the raw data into the form which can be accepted by the machine learning model as shown in Fig. 1.4. Initially the data are collected in an unstructured manner from different sources during the data collection, thus the data requires refining before involvement in the machine learning modeling. To provide quality data some approaches are required to be taken such as formatting of data, data cleaning (handle incomplete data due to noise like missing values), and sampling of data.

Therefore the conversion from inconsistent, incomplete, and error-prone data into an understandable, clean, and structured form allows the enhancement of the accuracy and efficiency of the ML model. As a result, the enhanced ML model might be able to provide precise and optimal results.

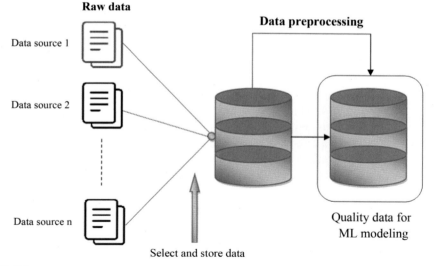

FIGURE 1.4

Data processing: data preprocessing.

8 CHAPTER 1 Machine learning architecture and framework

1. Data formatting: the importance of data formatting grows when data is acquired from various sources by different people. The first task for a data scientist is to standardize record formats. A specialist checks whether variables representing each attribute are recorded in the same way. Titles of products and services, prices, date formats, and addresses are examples of variables. The principle of data consistency also applies to attributes represented by numeric ranges.

2. Data cleaning: this set of procedures allows for the removal of noise and the fixing of inconsistencies in the data. A data scientist can fill in missing data using imputation techniques, e.g., substitute missing values with mean attributes. A specialist also detects outliers' observations that deviate significantly from the rest of distribution. If an outlier indicates erroneous data, a data scientist deletes or corrects them, if possible. This stage also includes removing incomplete and useless data objects.

3. Data sampling: big datasets require more time and computational power for analysis. If a dataset is too large, applying data sampling is the way to go. A data scientist uses this technique to select a smaller but representative data sample to build and run models much faster, and at the same time to produce accurate outcomes.

1.2.2.4 Transformation of data

This step involves the conversion of data into a form suitable for the ML model. For data transformation the following techniques are used:

1. Scaling of data: this is also known as the normalization of data. If attributes (features) of data are numeric then the scaling of the attributes becomes essential to put them into a common scale. The numeric attributes of data have different ranges, such as kilometers, meters, and millimeters, representing the data value. For example, scaling of an attribute for a minimum value may be in the range 0 to 10, and for the maximum value, it may be in-between 11 and 100.

2. Decomposition of data: it breaks the complex data into smaller part which makes it easier to understand the data patterns. Decomposition is a technique that overcomes the issue of the complex concept of attributes and breaks the complex features into simple subfeatures that makes the data more understandable and meaningful. Also, it guides the machine to where the new features can be added. Decomposition is mostly applicable for the analysis of time series data. Estimating a demand for goods per month for local vendors, a market analysis based on the data of big organizations to know the demand of goods per three months, or per six months, are examples of decomposition.

3. Aggregation of data: aggregation can be understood as the reverse process of decomposition. In aggregation, several features are combined into a single feature and whenever required all features can be explored. In other words, the aggregation represents the summarized form of the collected data which

can be used further for data analysis. This is the significant step of data transformation, where the quality and amount of data decide the accuracy of the data analysis. There are various applications where data aggregation can be used, which may include the finance sector, marketing plans, production-related decisions, and product pricing.

4. Feature engineering: this is a significant task in ML architecture in which the required features are selected and extracted from the data, which is relevant to the task or model to be developed, as shown in Fig. 1.5. The relevant features of data are further used to enhance the predictive efficiency ML algorithms. Therefore feature engineering should be done correctly, otherwise it will affect the overall development of the ML model. Feature engineering involves four subtasks: (1) feature selection: it deals with selecting the most useful and relevant features from the data; (2) feature extraction: selected features are extracted from the data for data modeling; (3) feature addition: existing features are added with new features selected and extracted from newly gathered data; and (4) feature filtering: irrelevant features are filtered out to make ML model more efficient and easy to handle.

1.2.3 Data modeling

Data modeling involves the selection of an appropriate algorithm which should be most adaptable for the system to address the issues in the problem statement [31]. It involves providing training to a ML algorithm to do predictions based on the available features, parameter tuning as per the business needs, and its validation on the sample data. The algorithms involved in this process are evolved through learning the environment and applying the training data set used in the learning process [32]. A trained model received after successful modeling is used for inference which allows the system to do predictions on new data inputs. The process of data modeling is shown in Fig. 1.6.

In the data modeling stage, various models are trained by the data scientist. The objective of this stage is to identify the model, whose prediction accuracy is better in comparison to others.

The data used for model training are categorized into two subsets. The first subset of the data is known as a training dataset which is used as the input to assist the ML algorithm during the training of the model. The input data is then processed by the ML algorithm which gives a model for predictive analysis on new data.

The training continues until we get the desired model. The training helps the model to improve its predictive hypothesis for new data. In other words, we can say that this makes the model able to predict the intended value from the new data. The training data can be labeled or unlabeled.

The labeled data has the value associated with it, while the unlabeled data has no predefined value.

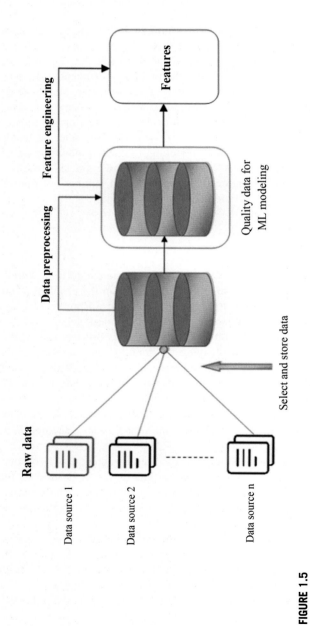

FIGURE 1.5

Data processing: feature engineering.

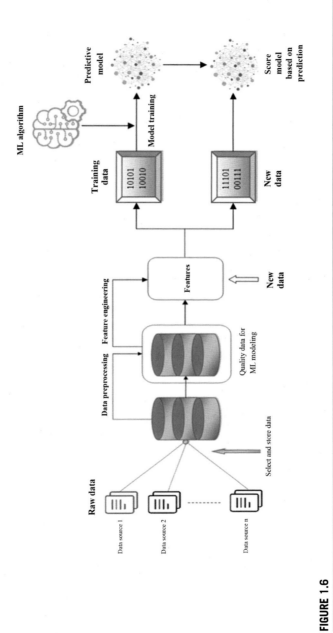

FIGURE 1.6

Data modeling: model training and scoring model based on new data.

12 CHAPTER 1 Machine learning architecture and framework

The second subset of data is known as test data. The test data is used to test the predictive hypothesis of the model which is created during the training. The overall objective of the data modeling is to develop such a model, which is able to do the predictive analysis of future data with high accuracy.

1.2.4 Execution (model evaluation)

The execution stage involves application, testing, and fine tuning of the algorithm (model) on test data set (unseen data). The objective of this stage is to retrieve the expected outcome from the machine and to optimize the system performance at the maximum level [33].

At this stage of ML architecture, the solution provided by the system is capable enough to explore and provide the required data for decision-making by the machine [34], as shown in Fig. 1.7.

1.2.5 Deployment

Deployment is the crucial stage which decides how the model will be deployed into the system for decision-making. At this stage the model actually is applied in a real scenario and also undergoes further processing. Further the output of the working model is applied as an instruction into the system for decision-making activities [35].

The output of the ML operations is directly applied to the business production where it plays a significant role in enabling the machine to take output-based expert decisions without dependency on other factors. The deployment of ML model is shown in Fig. 1.8.

1.3 Machine learning framework

The ML framework can be understood as a library, an interface, or a tool that helps working professionals to develop models of machine learning with ease without worrying about the underlying principles and complexities of the algorithms [36]. The framework provides the optimized and prebuilt components to build easy, meaningful, and quick ML models and other related tasks.

1.3.1 Features of ML framework

To choose the right framework, one should keep the following key features in mind [37] (https://hackernoon.com/top-10-machine-learning-frameworks-for-2019-h6120305j).

- Framework should be able to provide optimized performance.
- It should be easy and friendly to handle by the developer community.

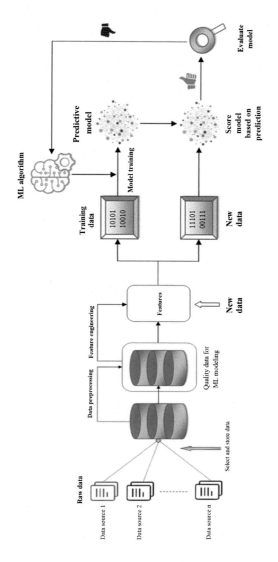

FIGURE 1.7

Execution: model evaluation.

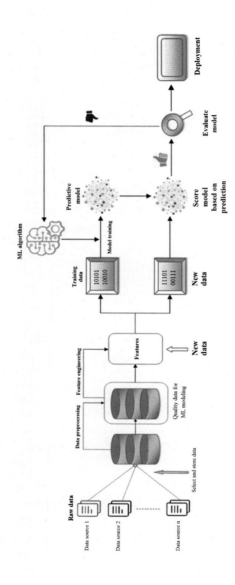

FIGURE 1.8

Deployment of ML model.

- The framework must be able to provide traditional ways of model development, and an easy and understandable coding style.
- It should be fully supported by parallelization for the distribution of computational processes.
- It must be supported by a strong and supporting user base and community.
- It should be out-of-box and avoid the black box approach.
- It should be able to reduce the complications of ML algorithms, making it more friendly and easily available to developers

1.3.2 Types of ML framework

ML frameworks have different categories based on their working principles which help developers to select them for their particular work. Before going into detail, a few insights are required to be acknowledged about the ML framework, which are as follows (https://analyticsindiamag.com/machine-learning-framework-10-need-know/):

- Few ML frameworks are focused on deep learning and neural networks.
- Few are oriented toward mathematical principles.
- Few are statistical analysis tools while others are linear algebra-based tools.

The five most popular ML frameworks have been discussed below based on their simplicity and developer-friendly environment.

1.3.2.1 TensorFlow

TenserFlow is an open source library specially designed to support and execute programs for dataflow and also covers a wide range of operations.

For the internal use of Google especially in the research and development activities, the Google Brain team developed this framework on November 9, 2015 under the open source license of Apache 2.0 [38,39].

Various machine-learning algorithms and tasks, such as classification, regression, and neural networks can be easily handled by TensorFlow. Both graphic processing units (GPUs) and central processing units (CPUs) support the running of this framework without any extra effort. Also, it can run on other platforms such as desktops, servers, and mobile devices [38,39].

This framework is very flexible and easily adaptable by the system and supports various versions and models as well. Due to these capabilities, it is very easy to migrate from nonautomatic to newer versions.

1.3.2.2 Amazon machine learning

Amazon ML is a very popular and robust service from Amazon Inc. It is a cloud-based service that makes it easy for developers with different skills to use all the ML techniques and to apply ML algorithms on real-world problems [40].

Amazon ML is supported by wizards and visualization tools that help to create different models of machine learning without needing to have the deep

16 CHAPTER 1 Machine learning architecture and framework

understanding of technical and theoretical concepts of the complex ML algorithms [41].

After the successful creation of your model, Amazon ML provides an easy and friendly environment for the developers to apply simple APIs. These APIs will be further applied to perform predictions for the application and also help to avoid the use of code generation by custom predictions.

1.3.2.3 Scikit-learn

In 2007 David Cournapeau introduced the Scikit-Learn which was a part of summer code project of Google [42].

It is a powerful and open-source python library to code machine-learning models. Libraries like Pandas, Numpy, IPython, SciPy, Matplotlib, and Sumpy are the building blocks of Scikit-Learn.

It also provides various functionalities such as regression, classification, clustering, preprocessing, and statistical modeling to solve real-world problems [29].

Scikit-Learn also offers some other functions which are very advanced and rarely supported by other libraries, such as ensemble, feature manipulation, outlier detection, and selection and validation of models [43].

Data mining and other various practical tasks can be easily handled by this framework. This framework also provides the required ease and speed to handle difficult tasks.

1.3.2.4 Apache mahout

This is an open-source tool of Apache foundation software which is specially designed to work and execute on machine-learning algorithms very quickly. It especially works as a framework for distributed linear algebra [44].

It also provides scalable performance for developing ML applications, especially in the field of classification, clustering, and collaborative filtering [45].

It provides an interactive environment for developers to create their own mathematical models. For general mathematical operations, Apache Mahout provides Java libraries.

1.3.2.5 Cognitive toolkit of microsoft

This is an open-source product of Microsoft for deep learning. It is specially designed for training purposes to solve complex problems, e.g., understanding the complex biological structure of the human brain. Also this tool can be used to develop various models, such as recurrent neural networks, feed forward DNNs, and convolutional neural networks [46].

This toolkit can be customized as per requirements, which makes it different from the other frameworks. A developer can easily choose algorithms, metrics, and networks as per their requirements. It allows neural networks (NNs) to use unstructured and huge data sets. The training time and architecture development support provided by this framework is very fast.

Also the back-ends provided by this toolkit are multimachine—multi-GPU [47].

The toolkit can be easily included in the library of C++, Python, and C#. It can also be used as a standalone tool for ML by using its BrainScript language. BrainScript language is a model description language [48].

1.4 Significance of machine learning in the healthcare system

In recent years, massive growth has been recorded in the worldwide population. Therefore, recording, maintaining, and analyzing the huge amount of medical data to provide useful information to the clinicians so they can diagnose and choose better remedial options is becoming a challenging task for the healthcare system. Specifically, undeveloped and developing countries are generally suffering from the lack of specialist doctors for certain diseases, unavailability of medical equipment, and fewer facilities for personalized treatment. In this context, ML can play a significant role in the detection and treatment of specific diseases. Also, it can ensure the best treatment for every patient on a personalized basis. ML can improve workflow efficiency in hospitals, drug manufacturing, and the development of medical equipment.

In recent years, ML has shown its ability in the healthcare system based on its inherent features to work with massive amounts of medical datasets in comparison to human potential. ML automatically processes the data and provides the required information which helps clinicians to make better planning and take appropriate decisions to take care of the patients. As a result, the doctors become able to reduce the cost of treatment and also ensure patient satisfaction. On another side, ML helps drug manufacturers to make new medicines, to understand the impact of medicine on patients, and to give effective suggestions for the improvement of medicines.

ML analyzes the complex medical datasets and presents them in a digital form. Further, it processes the data with its inbuilt efficient and effective computing mechanism. Various ML computing platforms are now available in the public domain at a reasonable price to do analysis on the issues related to the healthcare system. Some open source frameworks equipped with inbuilt libraries and tools are also available in the public domain.

Due to the growing market capital and pressure of analyzing the huge amounts of data, the healthcare system needs new features and dimensions with ML algorithms. Recently, a new ML algorithm has been introduced by Google to work for the detection of cancerous tumors through mammography images (https://www.mercurynews.com/2017/03/03/google-computers-trained-to-detect-cancer/).

To diagnose the symptoms of skin cancer, a deep-learning algorithm has been presented by the Stanford Artificial Intelligence Laboratory (https://news.stanford.edu/2017/01/25/artificial-intelligence-used-identify-skin-cancer/).

18   **CHAPTER 1** Machine learning architecture and framework

1.4.1 Machine-learning applications in the healthcare system

There are various applications of ML that have been widely available in the healthcare system, some of them are as follows:

1.4.1.1 Identification and diagnosis of disease

Due to population growth, current lifestyle, and growing pollution, the healthcare system is facing various challenges. These challenges may include improper handling of huge medical datasets and the shortage of automatic tools for disease identification, diagnosis, and cure. Many research laboratories are working together with medical professionals, clinicians, and drug manufacturing companies to provide the required solutions like diagnostic tools, and treatment kits for identification and diagnosis in the early stage of disease. Critical diseases like cancer, kidney failure, heart disease, and other single-gene inheritance, sickle cell anemia, and cystic fibrosis require real-time identification and diagnosis that help to provide proper medication on time.

If we look at the research initiatives taken for the identification and diagnosis of several diseases, it shows how various steps have been taken by different organizations to provide real-time solutions. To speed up the diagnosis of the tumor, an integrated cognitive computing and tumor sequencing (genome-based) has developed, named IBM Watson for Genomics (https://www.ibm.com/in-en/marketplace/watson-for-genomics).

Another biotechnology company "BERG," powered by AI, has developed an AI-based therapeutic treatment in various domains such as neurology, rare disease, and oncology (https://www.berghealth.com/).

1.4.1.2 Discovery and manufacturing of drugs

ML is applicable at every stage of drug discovery, starting from the initial stage, i.e., like the design of drug structure (chemical structure), to the middle stage, i.e., the validation of the drug, and to the final stages, such as clinical trials, manufacturing, and supply in the market.

ML can help the healthcare system in two ways: firstly, it can provide appropriate solutions to reduce the overall cost of drugs at the time of their introduction into the market.

Secondly, as compared to traditional techniques, it can accelerate the process of drug discovery and manufacturing, thus making it faster and as a result more cost-effective. The clinical application of ML can provide support to find the optimal solution for multifactorial disorders and genetic predisposition through precision medicine and next-generation sequencing techniques [49] (https://www.flatworldsolutions.com/healthcare/articles/top-10-applications-of-machine-learning-in-healthcare.php).

AtomNet is a deep learning neural network-based platform of "Atomwise Company," and has been specially developed for the design and discovery of drugs based on protein structure. This technique takes a day or two to learn millions of potential molecules, compared with the traditional techniques that take months. After learning protein molecules, the behavior of the medicine is analyzed in the patient through the simulation of the AtomNet. This makes it able to discover potential medicine to fight against various diseases (https://www.entrepreneur.com/article/341626). Another product of Alphabet Incorporation, which is the subsidiary company of Google, is named Deepmind and it is working in the same field, as discussed above, and huge progress has been made.

1.4.1.3 Diagnosis through medical imaging

In medical imaging, a huge amount of data can be processed through ML algorithms at very high speed. Medical images are vast datasets in healthcare systems that can be used for training to understand the minute facts of MRIs and CT scans by applying ML algorithms. A Microsoft product named InnerEye has been used to analyze images to produce images diagnostic tools. Companies such as Sophia and Enlitic have developed ML-based analysis tools to diagnose abnormalities in medical imaging reports. These tools work on all types of reports of medical images and analyze them with high accuracy in comparison to medical professionals.

For the early detection of breast cancer metastasis, Google has developed an assistance system named LYmph Node Assistant (LYNA). This system assists the pathologists and reduces their burden by generating reports with high accuracy. Another system has been developed by a joint effort by France, Germany, and the United States. This system is based on Convolutional Neural Network (CNN) to diagnose skin cancer with high accuracy (https://www.entrepreneur.com/article/341626).

1.5 Conclusion

At the current time, the world is evolving at a rapid pace and thus the needs of the people are also growing. In addition, we are in the era of the fourth industrial data revolution.

So, in order to retrieve useful and meaningful facts from big data and also to learn the different methods of interaction between the machine and humans efficient algorithms are required. These algorithms should be able to analyze all aspects of data and provide useful results for us to solve various purposes.

Machine learning has played a significant role in various domains of society, especially in healthcare, manufacturing, social media platforms, medicine, and many other sectors. Thus ML has become an important part of our lives.

Due to the exponential growth of data it is not possible to process and analyze the data by only increasing the computing power of the machine.

It needs some additional mechanism to examine the data deeply to provide fast and useful information. ML has shown its significance by providing information the way we need information. Some of the most useful ML examples are

CHAPTER 1 Machine learning architecture and framework

YouTube, Netflix, and Amazon Prime, platforms which recommend to you the videos, songs, movies, and much more based on your streaming habits. This is only possible due to the use of ML algorithms. Another example is the role of the Google search engine where you can get the appropriate result of your search based on your browsing habits.

In most sectors complex redundant tasks are becoming fully automated in place of manual processing in order to obtain accurate and steady results with high accuracy. All this is possible due to the power of ML algorithms.

References

[1] C.M. Bishop, Pattern recognition and machine learning, Springer, 2006.
[2] Expert System, (2019) <https://expertsystem.com/machine-learning-definition/>.
[3] C.E. Sapp, Preparing and architecting for machine learning, Gart. Tech. Prof. Advice (2017) 1−37.
[4] S. Goldman, Y. Zhou, (2000). Enhancing supervised learning with unlabeled data. In ICML (pp. 327−334).
[5] M. Hardt, E. Price, N. Srebro, (2016). Equality of opportunity in supervised learning. In Advances in neural information processing systems (pp. 3315−3323).
[6] A.K. Jain, J. Mao, K.M. Mohiuddin, Artificial neural networks: a tutorial, Computer 29 (3) (1996) 31−44.
[7] J.M. Zurada, Introduction to artificial neural systems, vol. 8, West, St. Paul, 1992.
[8] D.G. Kleinbaum, K. Dietz, M. Gail, M. Klein, M. Klein, Logistic regression, Springer-Verlag, New York, 2002.
[9] L. Wang (Ed.), Support vector machines: theory and applications, vol. 177, Springer Science & Business Media, 2005.
[10] G.E. Batista, M.C. Monard, A study of K-Nearest neighbour as an imputation method, His 87 (251-260) (2002) 48.
[11] I. Rish, (2001). An empirical study of the naive Bayes classifier. In IJCAI 2001 workshop on empirical methods in artificial intelligence (vol. 3, no. 22, pp. 41−46).
[12] A. Criminisi, J. Shotton, E. Konukoglu, Decision forests for classification, regression, density estimation, manifold learning and semi-supervised learning, Microsoft Res. Cambridge, Tech. Rep. 5 (6) (2011) 12.
[13] A. Criminisi, J. Shotton, E. Konukoglu, Decision forests: a unified framework for classification, regression, density estimation, manifold learning and semi-supervised learning, Comp. Graph. Vis. 7 (2−3) (2012) 81−227.
[14] L. Wasserman, J.D. Lafferty, (2008). Statistical analysis of semi-supervised regression. In advances in neural information processing systems (pp. 801−808).
[15] H.B. Barlow, Unsupervised learning, Neural Computation 1 (3) (1989) 295−311.
[16] M. Weber, M. Welling, P. Perona, (2000). Unsupervised learning of models for recognition. In European conference on computer vision (pp. 18−32). Springer, Berlin, Heidelberg.
[17] Z. Ghahramani, Unsupervised learning, Summer School on Machine Learning, Springer, Berlin, Heidelberg, 2003, pp. 72−112.

[18] T. Hastie, R. Tibshirani, J. Friedman, Unsupervised learning, The elements of statistical learning, Springer, New York, NY, 2009, pp. 485–585.

[19] M.A. Ranzato, F.J. Huang, Y.L. Boureau, Y. LeCun, (2007). Unsupervised learning of invariant feature hierarchies with applications to object recognition. In 2007 IEEE conference on computer vision and pattern recognition (pp. 1–8). IEEE.

[20] M. Caron, P. Bojanowski, A. Joulin, M. Douze, (2018). Deep clustering for unsupervised learning of visual features. In proceedings of the European conference on computer vision (ECCV) (pp. 132–149).

[21] P. Harrington, Machine learning in action, Manning Publications Co, 2012.

[22] R.S. Sutton, A.G. Barto, Introduction to reinforcement learning, vol. 135, MIT press, Cambridge, 1998.

[23] L.P. Kaelbling, M.L. Littman, A.W. Moore, Reinforcement learning: a survey, J. Artif. Intell. Res. 4 (1996) 237–285.

[24] S.B. Thrun, (1992). Efficient exploration in reinforcement learning.

[25] J. Attenberg, P. Melville, F. Provost, M. Saar-Tsechansky, Selective data acquisition for machine learning, Cost-sensitive Mach. Learn. (2011) 101.

[26] Z. Zheng, B. Padmanabhan, Selectively acquiring customer information: a new data acquisition problem and an active learning-based solution, Manag. Sci. 52 (5) (2006) 697–712.

[27] J. Kerr, P. Compton, (2003). Toward generic model-based object recognition by knowledge acquisition and machine learning. In proceedings of the eighteenth international joint conference on artificial intelligence (pp. 9–15).

[28] J. Qiu, Q. Wu, G. Ding, Y. Xu, S. Feng, A survey of machine learning for big data processing, J. Adv. Signal. Process. 2016 (1) (2016) 67.

[29] F. Pedregosa, G. Varoquaux, A. Gramfort, V. Michel, B. Thirion, O. Grisel, et al., Scikit-learn: machine learning in Python, J. Mach. Learn. Res. 12 (2011) 2825–2830.

[30] G. Manogaran, V. Vijayakumar, R. Varatharajan, P.M. Kumar, R. Sundarasekar, C. H. Hsu, Machine learning based big data processing framework for cancer diagnosis using hidden Markov model and GM clustering, Wirel. personal. Commun. 102 (3) (2018) 2099–2116.

[31] T.M. Mitchell, The discipline of machine learning, vol. 9, Carnegie Mellon University, School of Computer Science, Machine Learning Department, Pittsburgh, 2006.

[32] S. Sra, S. Nowozin, S.J. Wright (Eds.), Optimization for machine learning, MIT Press, 2012.

[33] E.R. Sparks, A. Talwalkar, D. Haas, M.J. Franklin, M.I. Jordan, T. Kraska, (2015). Automating model search for large scale machine learning. In proceedings of the sixth ACM symposium on Cloud Computing (pp. 368–380).

[34] Y. Reich, S.V. Barai, Evaluating machine learning models for engineering problems, Artif. Intell. Eng. 13 (3) (1999) 257–272.

[35] J. Boyan, D. Freitag, T. Joachims, (1996). A machine learning architecture for optimizing web search engines. In AAAI workshop on Internet based information systems (pp. 1–8).

[36] A. Menon, O. Tamuz, S. Gulwani, B. Lampson, A. Kalai, (2013). A machine learning framework for programming by example. In international conference on machine learning (pp. 187–195).

22 CHAPTER 1 Machine learning architecture and framework

[37] M. Feurer, A. Klein, K. Eggensperger, J. Springenberg, M. Blum, F. Hutter, (2015). Efficient and robust automated machine learning. In Advances in neural information processing systems (pp. 2962–2970).

[38] M. Abadi, A. Agarwal, P. Barham, E. Brevdo, Z. Chen, C. Citro, et al. (2016a). Tensorflow: large-scale machine learning on heterogeneous distributed systems. arXiv preprint arXiv:1603.04467.

[39] M. Abadi, P. Barham, J. Chen, Z. Chen, A. Davis, J. Dean, et al. (2016b). Tensorflow: a system for large-scale machine learning. In 12th {USENIX} symposium on operating systems design and implementation ({OSDI} 16) (pp. 265–283).

[40] A.S. Rathor, A. Agarwal, P. Dimri, Comparative study of machine learning approaches for Amazon reviews, Procedia Computer Sci. 132 (2018) 1552–1561.

[41] R. Shokri, M. Stronati, C. Song, V. Shmatikov, (2017). Membership inference attacks against machine learning models. In 2017 IEEE symposium on security and privacy (SP) (pp. 3–18). IEEE.

[42] R. Garreta, G. Moncecchi, Learning scikit-learn: machine learning in python, Packt Publishing Ltd, 2013.

[43] G. Varoquaux, L. Buitinck, G. Louppe, O. Grisel, F. Pedregosa, A. Mueller, Scikit-learn: machine learning without learning the machinery, GetMobile: Mob. Comput. Commun. 19 (1) (2015) 29–33.

[44] S. Owen, S. Owen, (2012). Mahout in action.

[45] D. Lyubimov, A. Palumbo, Apache mahout: beyond MapReduce, CreateSpace Independent Publishing Platform, 2016.

[46] K.V. Cherkasov, I.V. Gavrilova, E.V. Chernova, A.S. Dokolin, The use of open and machine vision technologies for development of gesture recognition intelligent systems, J. Phys.: Conf. Ser. 1015 (3) (2018) 032166. IOP Publishing.

[47] S. Pathak, P. He, W. Darling, (2017). Scalable deep document/sequence reasoning with cognitive toolkit. In proceedings of the 26th international conference on world wide web companion (pp. 931–934).

[48] M. Salvaris, D. Dean, W.H. Tok, Microsoft AI platform, Deep learning with Azure, Apress, Berkeley, CA, 2018, pp. 79–98.

[49] M.L. Gonzalez-Garay, The road from next-generation sequencing to personalized medicine, Personalized Med. 11 (5) (2014) 523–544.

CHAPTER

Machine learning in healthcare: review, opportunities and challenges

2

Anand Nayyar[1], Lata Gadhavi[2] and Noor Zaman[3]

[1]*Faculty of Information Technology, Graduate School, Duy Tan University, Da Nang, Vietnam*
[2]*IT Department, Government Polytechnic Gandhinagar, Gandhinagar, India*
[3]*School of Computer Science and Engineering SCE, Taylor's University, Subang, Jaya, Malaysia*

2.1 Introduction

Health is the most important aspect of human lives. During the healthcare process, a medical practitioner provides collected clinical data of each particular patient to diagnose the disease and to determine how to medicate the patient for the particular disease. Clinical data play a crucial role in addressing the health issue, and machine learning techniques can help to diagnosis the disease. Machine learning plays a crucial role in the healthcare process for disease diagnosis, decision support system, prognosis, and the perfect treatment plan for the detected disease. Machine learning techniques can be used for health-related usage, like diagnosis of disease, identification of prescriptions etc. Machine learning architecture and methods will help to enhance the effectiveness, accuracy, and quality in the medical field, and facilitate new approaches for disease diagnosis and treatment plans. In this chapter, we review the latest research in the field of healthcare using machine learning and present the futuristic opportunities to build the machine learning-enabled system. This chapter will introduce a literature review that will analyze the current work and prepare for future work concerning machine learning and healthcare.

The integration of healthcare and technology indicates some characteristics such as;

1. Adoption of smart systems to handle health issues of human beings.
2. Improvement in quality of life of human beings using intelligent and smart systems.
3. Distribution of human workload to machine.
4. Reducing the high expenditure on healthcare.

Machine Learning and the Internet of Medical Things in Healthcare. DOI: https://doi.org/10.1016/B978-0-12-821229-5.00011-2
Copyright © 2021 Elsevier Inc. All rights reserved.

CHAPTER 2 Machine learning in healthcare

By including artificial intelligence in the healthcare applications, the healthcare system can become a smart system. Using machine learning in healthcare applications, disease diagnosis can optimize the time of medical professionals. To monitor and observe the effectiveness of treatment in the healthcare field, machine learning applications can be used for diagnosis, prognosis, and to provide the perfect treatment plan for the detected disease.

To reduce the cost of healthcare and improve personalized health, the healthcare field faces challenges in electronic data management, electronic diagnosis of disease, and integration of data with the healthcare system. Machine learning provides a variety of techniques, methodologies, and tools which can solve the addressed challenges. This chapter details the tools and technology of machine learning for use in the healthcare domain to alleviate the faced challenges. Data is a significant resource in machine learning technology. With the quality and quantity of data, machine learning can improve its efficiency and prediction progression. Machine learning has been largely applied to solve the issues of healthcare. It can be used to identify the treatment plan for a patient, provide advice to patient, predictions, and disease summary based on the found symptoms with more accuracy. The algorithm of machine learning can provide lifestyle advice to a patient based on the present medical condition and history. The models of machine learning can be trained to predict outcomes. Machine learning technology has lots of scope for research in healthcare. Machine learning in healthcare is used to identify the discrepancy of a patient to inform a person about medication. This chapter will cover various issues, and challenges faced by practitioners in the last few years. A brief history of the field of healthcare and artificial intelligence will be provided, which will serve as a motivation and source of ideas to the researchers.

To meet the above objectives the remaining part of this chapter is organized into five sections. Section 2.1 represents an overview of machine learning with healthcare. Section 2.2 shows domain analysis of field machine learning and healthcare. In section 2.3, identified challenges in healthcare and machine learning are presented. Finally, section 2.4 presents a summary of the chapter.

2.1.1 Machine learning in a nutshell

The first question that arises when we hear about machine learning terminology is: what is it? It seems like a magic wand that everyone is fascinated with. Nowadays, many technology giants are integrating their application and startups with machine learning techniques at their core. The algorithms of machine learning are trained to learn from past experiences to predict the future and making decisions accordingly. A task is given to develop a game of rock, paper, and scissors that can play against a human. Imagine what it would be to create an application which understanding the combinations, sensing, and differentiating between hand gestures and so on. The task grows even more complex when we add variables like recognizing the opponent, drawing up a strategy, etc. Machine learning can help to perform the above task.

In traditional programming, outputs are generated based on the given guidelines. However, in machine learning, programmers provide the computer with a ton of answers with data and the computer will generate the guidelines for them. In the rock, paper, and scissors example, if we provide the computer with hundreds of images of hand formations of rock, paper, and scissors and the computer learns the patterns that allows it to make accurate guesses on its own, it means we have achieved machine learning. Technically, machine learning is the next immense significance of big data analysis. It is an evolving branch of computational algorithms which are designed to replicate human intelligence by learning from experience [1].

2.1.2 Machine learning techniques and applications

Exponential growth of data needs a system which can handle this huge load of data. Machine learning is the modern way to perceive and handle the information for accurate predictions in the future. Machine learning algorithms use the patterns enclosed in the training data to accomplish classification and make future predictions. Machine learning algorithms can be classified into three types as shown in Fig. 2.1;

- Supervised learning
- Unsupervised learning
- Reinforcement learning

2.1.2.1 Supervised learning

In supervised learning, a labeled model will be trained with the mapping of input and output. Supervised learning uses machine learning technology to train the mapped function from input to the output. It provides the machine with known quantities to support future guesses made by the algorithm [2]. Supervised learning, as the name implies, involves the presence of a supervisor to supervise the task. In supervised learning we train the machine using the labeled data so the machine will prepare a new set of data for the next prediction. For example, you are given set of different shapes in one basket. So, the first step is to train the machine to identify the different kinds of shapes, like triangle, circle, hexagon, and square as shown in Fig. 2.2. The machine will learn from the previous data and it applies the algorithm to classify the different shapes. Supervised learning can be split into classification and regression. In classification, input data is mapped to output labels and in regression input data is mapped to a continuous output [3].

26 CHAPTER 2 Machine learning in healthcare

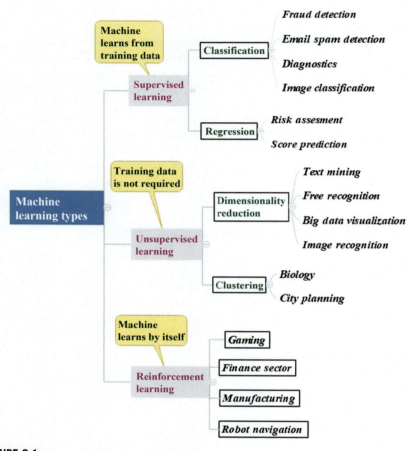

FIGURE 2.1

Types of machine learning techniques.

2.1.2.2 Unsupervised learning

There is no empirical data involved with the unsupervised learning technique unlike supervised learning. The computation is done solely on the input provided to the system with no prior labels of classification made [4]. In this technique, the group are unsorted according to the similar patterns without any prior training of data. No supervisor will be provided to the machine for training. For example, if different images of shapes are given to the machine which have not been seen ever by a machine then the machine will categorize them according to the similarities, differences, and patterns, as shown in the Fig. 2.3. It is classified into two categories like clustering and dimensionality reduction.

2.1 Introduction   27

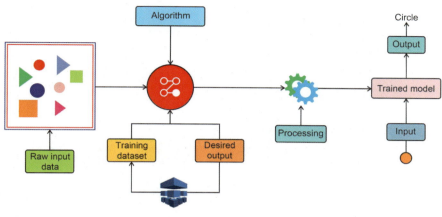

FIGURE 2.2

Supervised learning.

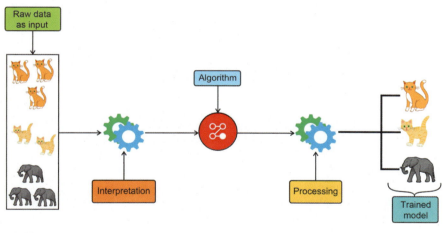

FIGURE 2.3

Unsupervised learning.

2.1.2.3 Reinforcement learning

This is an idea where a reward system is used to help the system to learn from its past behaviors [4]. For example, in a game of pong, when the system makes a bad move or misses the "ball," it would receive no reward. On the flipside, if the system makes the right move, it gains a reward, as presented in Fig. 2.4.

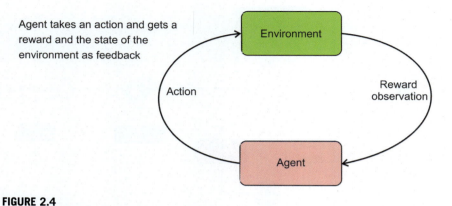

FIGURE 2.4

Reinforcement learning.

This type of "good—bad" reward scheme helps the algorithm to learn what it is supposed to do.

2.1.3 Desired features of machine learning

Based on the potential of the technology, we shall discuss some of the features that we desire in an ideal machine learning algorithm.

1. Accuracy

    Accuracy is the essential part of the algorithm implementation process, although it depends on the data being fed in to the system. Faulty data can cause inaccurate results. The attributes used in the learning process may be redundant or correlated, which may lead to degrading of the accuracy of prediction [5].
2. Speed

    The primary advantage of a machine algorithm is to perform tasks which require human level intelligence at the speed of a computer program. There is a myth that a good machine learning algorithm is the fastest one. In fact, a good algorithm strikes a unique balance between speed, accuracy, and efficiency. There is an apparent trade-off between speed and accuracy. Researchers are now working on algorithms which minimize the trade-off [6].
3. Efficiency

    A machine learning system should achieve maximum productivity with the minimum amount of wasted resources. When a system is presented with a large dataset, the computation requirement goes up exponentially. Current machine learning algorithms are performance driven—the focus is on the predictive/classification accuracy, based on empirical data fed to the system [7].

4. Unbiased

As mentioned earlier, a machine learning system is only as good as the data that goes into training it. If public systems are trained on data that is incomplete or faulty, it can cause serious problems. For example, a face recognition system was trained on a dataset which had images of primarily Caucasian subjects. Later in accuracy tests, the system failed to recognize individuals of color [8]. Therefore it is imperative to use a complete dataset to train an ML algorithm and they should be extensively tested before being used as public systems.

2.1.4 How machine learning works?

There is a distinct difference between workings of traditional computational algorithms and machine learning algorithms (Fig. 2.5).

As presented in Fig. 2.6, in a machine learning system the computation of the algorithm is driven by data and examples of answers provided to the algorithm. The goal here is to make the machine recognize the pattern of the answers provided to it, match it with relevant data, and try to come up with accurate predictions on its own after its training period. The internal workings of a machine learning algorithm can be classified into the four categories:

1. **Supervised learning**: the process of algorithm learning from the training dataset can be thought of as an instructor supervising the process of learning. Here, the instructor is aware of the correct answer and provides the algorithm with a correction when it makes a mistake. Hence, the learning process of the machine is "supervised."
2. **Semi-supervised learning**: the process of algorithm learning from a mixed dataset. Here, the data provided for training purposes has some labeled data and largely unlabeled data. This approach deals with scenarios where the dataset is not complete. The idea is to apply techniques of both supervised and unsupervised learning to achieve accurate predictions/classifications.

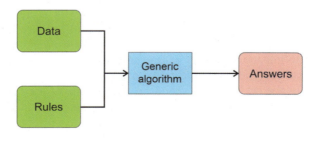

FIGURE 2.5

Normal algorithm flow.

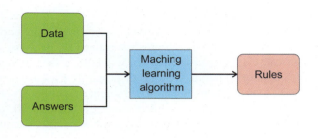

FIGURE 2.6

Machine learning algorithm flow.

3. **Unsupervised learning**: the process of algorithm learning from a dataset which contains no labeled data. The system is responsible for finding the correct pattern in the dataset.
4. **Reinforcement learning**: the process of algorithm learning by gaining feedback on all of the actions made by the system. In this type of learning process, the system is fed back its past action along with a "good" or "bad" score. If every good score adds $+1$ to the total and a bad score subtracts -1 from it, the goal is to keep the score as high as possible.

2.1.5 Why machine learning for healthcare?

Machine learning is omnipresent technology used for various fields like finance, healthcare, security, and in education. The healthcare field is a prominent research domain with fast technological advancement in which the amount of data is increasing day by day. To handle the large volume of data, advanced technology is required. In healthcare systems, vast numbers of patients inquire about the different treatments using various methods. Machine learning provides excellent disease diagnosis capabilities. Many of the current practices involved in manual processes of healthcare systems are tedious and slow. The healthcare field can utilize the potential of machine learning and improve services drastically. Medical practitioners can make better decisions about patient diagnosis and treatment options, while understanding the possible outcomes with more information. In healthcare, machine learning derives its value due to the ability to process huge data sets beyond the scope of human capability, and then reliably convert analysis of that data into important insights which can aid physicians in planning and providing care, ultimately leading to better outcomes, lower costs, and increased satisfaction [8]. In the healthcare domain, the significance of the machine learning has increased hugely. We accentuate that algorithms or methods developed using smart technology should not alter medical staff, but to some extent be used to optimize the medical workflow.

IoT devices have been used to track every movement and every moment of daily lives. Fitbits, smart watches, and sensors sense and generate surplus data everyday. This abundant data needs to be processed everyday so that actual health benefits could be ripped out of it. Machine learning can help crunch all these numbers and even go as far as to create personalized healthcare and treatment plans and also help patients to follow up their appointments by forming customized schedules.

Being provided with this plethora of data for healthcare has its own merits and demerits [9]. The ability of wearable devices and sensors to continuously track physiologic parameters can provide an overall patient care strategy for improving outcomes and lower healthcare costs in cardiac patients with heart failure [10]. The majority of medical data is unstructured data in the form of various different notes, reports, discharge summaries, images, and audio and video recordings.

Considering the risk of chronic diseases, it is much too difficult to quantify and characterize a communication between the healthcare provider and the patient. For two patients suffering a similar fever, the doctor, the conversation with the doctor, and overall diagnosis may be very different as per the interpretations of the doctor and the way the patient explains the scenario [9]. Hence, machine learning should serve as a boosting platform that helps in organizing all the unstructured data and even efficiently analyze the structured data.

Various studies and researches are being conducted in healthcare. If they are infused with machine learning to understand the patterns produced out of the experiments and behaviors of various systems, more insightful theories and results could be drawn. Machine learning can be applied to support, diagnose, and direct patients to accurate treatment all while keeping costs down by keeping patients out of expensive, time-intensive emergency care centers [11]. Machine learning cannot only help in the diagnosis of the diseases but it can also help to predict the risk of the disease. Using the perks of supervised machine learning, models can be trained to predict the risk of the diseases. These models could be based on the data such as age, height, weight, gender, and living habits (such as exercising or not, drinking or not) and others relevant to specific case data which might again be structured or unstructured.

The risk prediction accuracy would depend on the accuracy and timeliness of the data being provided and the type of disease risk being predicted. Given that the medical practices keep on changing over a period of time, machine learning could help improve the changes and formulate tactics for a smoother transition. Machine learning cannot only be helpful by its direct implementation in medicine but also be helpful by its other world features such as image optimization by colorizing and restoring, or simulating the human bodily systems to generate case specific and condition specific data. The increasing ability of data to be transformed into knowledge should rattle the medical domain [11]. Machine learning shall be responsible for improving the prognosis, decreasing much of the work of medical practitioners of consuming the raw data and drawing conclusions out of them, increasing the diagnostic accuracy, and predicting the risk of diseases accurately. Machine learning shall become an indispensable tool for medical practitioners as the complexity of patients' conditions and diseases increases [10].

2.2 Analysis of domain

Data are being increasingly generated as compared to the earlier times due to the sensing devices and modules. This has resulted in development to interact with and extract meaningful information from the generated data [12]. There are many hidden patterns in the data which are not always found by the human brain when interpreting the data. Healthcare has always been one of the most transforming divisions in human history. Humans have progressed a lot in the field of medical healthcare [9]. The healthcare industry is made up of preventive, diagnostic, remedial, and therapeutic sectors. Technological advancements have made significant progress in transforming everyday healthcare. Everyday appointments with the doctors and follow-ups with treatment can be delivered more efficiently. Using microchips and smart monitoring health devices, patients can be monitored more effectively and the patients will have to approach the doctors only if it is found necessary through the readings obtained from these monitoring devices. These technologies may not be able to completely nullify the human intervention required while treating the patients, however, it can make the whole process much more efficient and reduce the economical expenditure spent to improve the healthcare industry. Healthcare is a field which can progress much more significantly if the amount of data being provided is increased and availability is made easier. Machine learning is based on the availability of data on patients and can significantly improve services provided to them.

A study by the Institute on Medicine (IMO) in 2012 concluded that computing will play a huge part in advancing medicine [13].

2.2.1 Background and related works

In recent years, the research toward finding solutions in healthcare using machine learning techniques has increased at a steady pace. Using computation to automate basics tasks has been a long-standing integration challenge which was done with relative ease.

Before the advent of cheap powerful computing and cloud technologies that easily facilitate machine learning models, researchers worked extensively on rules-based systems. These kinds of systems use predefined rules, they behave predictably, and are meant to be used as a solution for reducing clinical errors in a medical process [14]. Despite the nature of their purpose, rules-based systems have caused problems in tests and they have been proven to be unreliable to handle the magnitude of data and rules that come up while managing a healthcare system [15].

There exist systems that generate automated notifications and patient care instructions based on historical data to increase self-care and improve clinical management in medical institutions [16]. Research has also been done to detect cardiac arrhythmia in e-healthcare systems using a predefined rules-based approach [17] and creating wireless body sensor networks for patients [18].

2.2.2 Integration scenarios of ML and Healthcare

Healthcare using machine learning can prove to be a huge leap forward in providing quality medical services to the general population. The integration of healthcare and technology indicates some characteristics such as:

1. Adoption of smart systems to handle health issues of human beings.
2. Improvement in quality of life of human beings using intelligent and smart systems.
3. Distribution of human workload to machine.
4. Reducing the high expenditure on healthcare.

By including artificial intelligence in the healthcare applications, the healthcare system can become a smart system. Using machine learning in healthcare applications, disease diagnosis can optimize the time of medical professionals. To monitor and observe the effectiveness of treatment in the healthcare field, machine learning applications can be used for diagnosis, prognosis, and perfect treatment plans for the detected disease. Machine learning models can be used to improve or provide:

- Better diagnosis from medical images
- Treatment options based patient data
- Chances of catching a particular disease after being admitted
- Prediction of zoonotic diseases reservoirs
- Drug discovery
- Automated surgeries
- Automatic personalized medicines and treatments

2.2.3 Existing machine learning applications for healthcare

As a part of digital era, the latest innovation is being carried out in machine learning. This intelligent technology is being used in different industries and academics. In the real world, there are many machine learning applications being used in fields like retail, healthcare, social media, e-commerce, manufacturing, transport, energy, and utilities. As presented in Fig. 2.7. Machine Learning giants like Google LLC, who have been exploring the technology and have the means to execute radical solutions on a large scale, have already started to roll-out machine learning-based systems into the real world. The company launched a ML powered project to detect diabetes-related eye diseases in India [19].

34 CHAPTER 2 Machine learning in healthcare

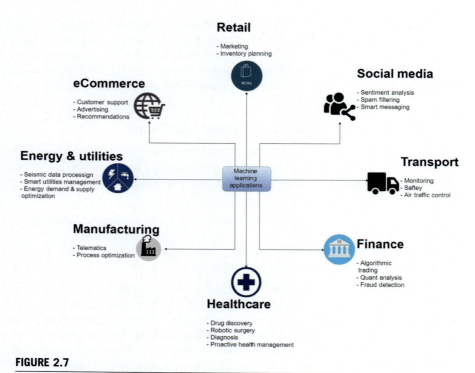

FIGURE 2.7

Machine learning applications.

Apart from Google, machine learning-based healthcare platforms are getting increasingly more attention. There are community projects for healthcare, i.e., a software platform which seeks to shine light on "increasing the national adoption of machine learning in healthcare." Researchers have also applied machine learning to classify cancer types based on gene selection from microarray data [20]. Apart from these, research has also been made to detect skin lesions [21] predict occurrences of sepsis [22], and to provide radiation treatment recommendations that are tailored to particular patients with minimal damage to nearby organs [23].

1. Retail

Retail is the economic base of how consumers purchase goods. At retail stores, virtual or otherwise, everything from demand of products to the location of the physical store account for the profit margin that the store will gain. Some factors that machine learning is improving in the case of retail are:

a. **Location:** for a physical retail store, it is important to choose an appropriate location to place the store. This is because the location and the demographics surrounding it play an important role in such businesses [24].

b. Sales: Sales generate revenue for the store. A high demand for an item will result in an increase of sales for that item. For example, let's say that for some reason the demand of apples skyrockets. In this case, the stores that sell apples will face a problem with the supply they have and will not be able to meet the demand. To prevent this, machine learning could help store owners to predict sales, they could stock themselves accordingly. This also prevents stores from buying surplus amounts of perishable items that might be wasted if not sold [25].

c. Inventory: all the stores need accurate inventories. An inventory tells the store owners about how they can advertise, when to order more items, and what not to buy again. Automated inventory systems that conduct accurate inventories using text and visual data are being developed to make this process efficient and easy [26].

2. e-Commerce

Every commercial trade that takes place on the internet falls under the category of e-commerce. With the advent of the internet and increase in usage, we have discovered a new way to trade and advertise products. Even traditional companies have moved out of brick-and- mortar retail stores toward a more online approach.

This ease of access brings problems of its own and machine learning can help to solve them. For example, using data mining we can detect fraudulent entities and purge them from the application environment [27]. Let's say that someone is trying to hoard products on an e-commerce website. We can use this system to detect their activities and block them from causing more harm.

Another area that can be improved is pricing. Prices play a huge role in a customer's decision to purchase a product. If the prices of all the items in an online store are controlled manually, it can prove to be a tedious task to change them as the market moves. A smart system uses a dynamic pricing model to set and change prices and take a huge load off the maintainers [28].

3. Energy and utilities

Energy is what makes everything work. Humans have learned to harvest energy from different sources to power man-made systems. In recent times, we have learnt the hard way that energy is not available in an unlimited quantity. Hence, it is imperative that we use, harvest, and manage these resources smartly.

Data centers are the by-product of the information age. All the data we collect need to be stored somewhere. The problem is that data centers take up a lot of energy. One estimate puts the global data center energy consumption per annum at 3% of all the electricity produced, almost 40 times the consumption of the entirety of the United Kingdom [29].

Apart from that, machine learning is also being used to estimate energy performance in residential buildings [30], improve the efficiency of underwater sensors by using smart routing protocols [31], and to study particle behavior and physics [32,33].

36 CHAPTER 2 Machine learning in healthcare

4. Manufacturing

Continuous and steady improvement is essential to manufacturing entities and necessitates flat and flexible organizations, lifelong learning of employees on the one side, and information and material processing systems with adaptive, learning abilities on the other side [34]. To address this issue, research has been done to improve the state of manufacturing using machine learning techniques [35].

5. Healthcare

Industries across the world are adopting machine learning. By the mid-2020s, the markets for smart solutions for the healthcare industry are predicted to be valued at US$ 34 billion, taking a huge 34% leap from current numbers.

Machine learning powering computer systems and detecting threats, such as like drug diversion and identity theft, running drug test simulations, predicting disease, and so much more, can improve the healthcare landscape on a huge scale.

Here are some areas where machine learning is vastly improving healthcare:

- Disease prediction [36]
- Informatics [37]
- Patient records management [38]
- Healthcare delivery [39]

6. Finance

Modern advancements in analytical techniques and computation, along with the availability of applications-oriented fields like Big Data, have made things like financial analysis, investment management, and trading possible for machines. Computers are the best at crunching numbers accurately. If we combine this power with the ability to make smart decisions and handle finances, the possibilities are limitless.

Option price is the amount per share that an option buyer pays to the seller. Machine learning is making the process easier [40]. Financial transactions can be made completely secure in a trustless environment by using blockchain and machine learning together [41].

7. Transport

The entire global economy depends on the transportation of materials and goods. Generally, only a few percentages of goods produced in regions are consumed there. Most of the time, these products are shipped to facilities or directly to the end consumers. Therefore the creation of systems to assist in this process that can take out the human error factor is something to be desired [42].

Think of a scenario where truck drivers spend hours on the road, working without rest. This is the reality of the situation now and it's causing loss of life and resources. Self-driving transport vehicles can put this loss to an end. Data mining and classification can be used to generate a safety map for vehicles to follow. Drones can also be used to deliver smaller payloads and companies like Google and Amazon are already experimenting in this direction.

Machine learning techniques are also being applied toward the determination of road suitability for the transportation of dangerous substances like chemicals, radioactive elements, and weapons [43].

8. Social media

Increasingly, we are seeing more people online on average than ever before. These numbers are huge—there are nearly 8 billion people in the world and of those about 3.5 billion are already online, according to World Development Indicators. This means that one in every three people is using social media on a regular basis.

People now tend to think and express online and all this activity generates an unprecedented amount of data. Machine learning with the power of computational hardware and data analytics abilities can change the landscape in the following areas:

a. Sentiment analysis: contextual mining of text, images, and even emojis which can gain subjective information on individual users. This can help producers and brands understand the demographic and its general mood toward some idea or product [44].
b. Behavior studies: the human mind, its behavior, and the decision-making process form a very complex system. Creating systems that can cater to the larger population requires research into the mind of that demographic and their behavior. Mining the data and projecting it as behavior of a certain region is indeed a really helpful process [45].
c. Background analysis: because the majority of us are online now, there is a lot of data of our past activity on the internet. Companies, law enforcement, and governments can use this data to gain an insight into the personality of a person [46].
d. Curbing bullying: cyber bullying and threats run rampant online. People tend to vent out verbally on some entity that does not conform to their world view. There are projects, like Perspective API, which can clear and curb toxicity from a social forum, keeping it clean.

2.3 Perspective of disease diagnosis using machine learning

The scope for disease diagnosis using machine learning has proven to be simple to achieve. At the time of writing this chapter, several researchers have published related work. Generally, most of the methods suggested either using artificial

CHAPTER 2 Machine learning in healthcare

neural networks or support vector machines. The following is the list of diseases that have been implemented for ML diagnosis:

- Alzheimer's disease

    Alzheimer's disease cannot be identified by any specific test. It can be diagnosed based on the clinical history of the patient, neuropsychological tests, and electroencephalography. So, new approaches are required to enable accurate diagnosis and get the accurate results [47,48]. Machine learning applications gained ample attention to detect and classify Alzheimer's disease (AD). Machine learning is an emerging area in which researcher may use raw neuroimaging data to identify features that attract attention in the field medical imaging analysis.

- Thyroid Disease

    The thyroid is situated in the lower region of the neck under the layers of skin and muscle which can be identified in women and men [49]. It is essential to generate an accurate tool for malignancy risk identification to improve the survival rate of patients. Moreover, early identification of the symptoms of thyroid disorders can increase the survival rate, thereby initiating the treatment at first stage. Therefore the diagnostic process and treatment are required to use machine learning technology to get accurate classification of thyroid nodules.

- Heart diseases

    The crucial disease in the middle or old age is heart disease which leads to the maximum complications. It is difficult to identify the heart disease symptoms manually [50]. However, machine learning is a useful technology to predict the symptoms of heart disease for the diagnosis of disease and treatment plans.

- Hepatitis

    Machine learning algorithms have shown remarkable attention in health care field in recent era. It is possible to identify powerful classifications for getting the effective information from the correlated and imbalanced medical datasets [51]. A simple example would be an individual visiting a healthcare facility complaining about chest pains. The ML model might search to look for the top five reasons for chest pains, declaring that they are healthy [52]. This approach can fail for all kinds of people who face uncommon diseases. We are far away from creating a system that can accurately diagnose all diseases but the research is moving in the right direction.

2.3.1 Future perspective to enhance healthcare system using machine learning

With the issue of machine learning in public healthcare system, the future seems to be bright. Machine learning is bringing a paradigm shift to healthcare, powered by the increasing availability of healthcare data and the rapid progress of analytics techniques. With computation costs and hardware sizes declining, machine learning systems are turning into mobile devices. Wireless personal health assistants with basic disease diagnostic capabilities are being developed by tech companies. Apart from neural networks and support vector machines, different machine learning structures are being used to overcome medical challenges. As we move toward the future, machine learning and healthcare can be combined with technologies like:

- **Internet of Things**: remote injections, automatic medicine delivery.
- **Blockchain**: secure and decentralized transfer of private data.
- **3D printing:** creating cheap drugs and equipment.

2.3.1.1 Challenges and risks

While the untapped reserve of innovations to be made in healthcare is huge, we also have to keep in mind the delicacy of the matter at hand. In real life, healthcare is not a joke and machine learning systems can cause serious damage if not taken seriously.

In a report on AI in healthcare [53], the authors mentioned:

"Development of AI in healthcare using the application of machine learning is an ample area of research, but the rapid pace of change, diversity of different techniques and multiplicity of tuning parameters make it difficult to get a clear picture of how accurate these systems might be in clinical practice or how reproducible they are in different clinical contexts"

[53].

The following list describes challenges and risks involved using a machine learning system to supplement a healthcare process as shown in Fig. 2.8.

1. **Data**

    Data as an input to the system can be in many forms. It is crucial that all the data being fed to the algorithm is being gathered from a vetted source. As a general rule, one must always assume that data will be messy and incomplete. The first challenge faced by a system that uses machine learning will be to clean up the data and still generate useful predictions [54].

2. **Black-box decision-making**

    Some methods to achieve machine learning, like neural networks, contain a hidden or abstract layer(s) in the process. It means a part of decision-making and rationale of the machine's process cannot be seen clearly. The problem has been realized most often by researchers working with ML algorithms which require image analysis [21].

    A black-box ML algorithm that outputs the identity of an individual at risk can be significantly less helpful than an open algorithm which can specify why they are at risk and what makes the system come to a particular decision.

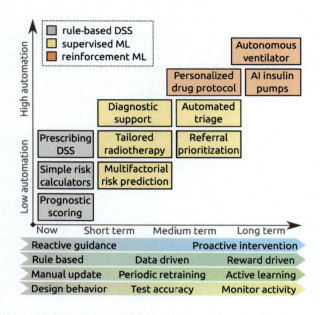

FIGURE 2.8

Issues with ML in healthcare [53].

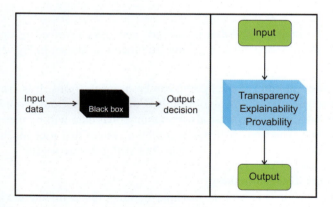

FIGURE 2.9

Black box versus transparent decision-making.

A system's ability to share its process openly can help the researchers or developers to find bugs easily. A black-box and transparent decision-making systems are compared in Fig. 2.9.

3. **Automation complacency**

It may happen that the physicians start to rely on the model so critical decision-making tasks are delegated to the machine. It can pose a catastrophic harm to the medical process. Automation bias [55] presented issues and models which cannot be trusted completely with the entire medical process.

4. **False positive and false negative results**

In a real-world scenario, a clinician might not understand the motivation of a machine learning system for outputting some results. This sort of oversight into the inner workings can influence the medical process through a false positive or a false negative result. A study in disease prediction using machine learning evaluated the accuracy of a system based on a calculation involving FPs and FNs generated by the system [56].

2.4 Conclusions

The research and development of machine learning model is a viable field but not without its problems. Rapid changes in the technologies and treatment methods make it difficult to track actual progress. Along with that, we cannot predict how these systems would behave in different real-world scenarios. The focus should also eventually be shifted from research to deployment of these algorithms. The research should be conducted with rigorous oversight.

It is hard to imagine a future without machine learning in healthcare. The question is: how fast can we get there and at what cost? Haphazard attempts at creating generalized solutions will only generate more problems. Hence, a step-by-step approach is necessary for the slow but sure development of machine learning systems. In this study, we presented an evaluation and comparison of well-known machine learning algorithms and their methodologies. This chapter also presented the potential of machine learning technology being used for medical outcome predictions. We reviewed the motivation of using machine learning in healthcare for various diseases. In this chapter, we indicated that there are many opportunities for machine learning researchers to work in the healthcare field to improve the current status of the medical system. Machine learning is playing an important role in the field of healthcare. It is possible to make more accurate systems for disease diagnosis and treatment plans of the patient using machine learning.

References

[1] M. Feng, G. Valdes, N. Dixit, T.D. Solberg, Machine learning in radiation oncology: opportunities, requirements, and needs, Front. Oncol. 8 (2018) 110.

[2] S.B. Kotsiantis, Supervised machine learning: a review of classification techniques. In Proceedings of the 2007 conference on emerging artificial intelligence applications in

computer engineering: real word AI systems with applications in eHealth, HCI, information retrieval and pervasive technologies, (2007), IOS Press, NLD, pp. 3−24.

[3] M.W. Sholom, K. Ioannis, An empirical comparison of pattern recognition, neural nets, and machine learning classification methods. In Proceedings of the 11th international joint conference on artificial intelligence - Volume 1 (IJCAI'89), (1989) Morgan Kaufmann Publishers Inc., San Francisco, CA, pp. 781−787.

[4] S. Nashif, M.R. Raihan, M.R. Islam, M.H. Imam, Heart disease detection by using machine learning algorithms and a real-time cardiovascular health monitoring system, World J. Eng. Technol. 6 (2018) 854−873.

[5] T. Howley, M.G. Madden, M.L. O Connell, A.G. Ryder, The effect of principal component analysis on machine learning accuracy with high dimensional spectral data. In Proceedings of the international conference on innovative techniques and applications of artificial intelligence, (2005), Springer, London, pp. 209−222.

[6] J. Huang, V. Rathod, C. Sun, M. Zhu, A. Korattikara, A. Fathi, et al., Speed/accuracy trade-offs for modern convolutional object detectors. In proceedings of the IEEE conference on computer vision and pattern recognition, (2017), pp. 7310−7311.

[7] O.Y. Al-Jarrah, P.D. Yoo, S. Muhaidat, G.K. Karagiannidis, K. Taha, Efficient machine learning for big data: a review, Big Data Res. 2 (3) (2015) 87−93.

[8] A. Golande, T.P. Kumar, Heart disease prediction using effective machine learning techniques, Int. J. Recent. Technol. Eng. 8 (1) (2019) 944−950.

[9] R. Bhardwaj, A.R. Nambiar, D. Dutta, A study of machine learning in healthcare. In 2017 IEEE 41st annual computer software and applications conference (COMPSAC), IEEE, (2017), vol. 2, pp. 236−241.

[10] S.R. Steinhubl, E.J. Topol, Moving from digitalization to digitization in cardiovascular care: why is it important, and what could it mean for patients and providers? J. Am. Coll. Cardiol. 66 (13) (2015) 1489−1496.

[11] M.A. Ahmad, C. Eckert, A. Teredesai, Interpretable machine learning in healthcare. In Proceedings of the 2018 ACM international conference on bioinformatics, computational biology, and health informatics, (2018), ACM, pp. 559−560.

[12] M. Mozaffari-Kermani, S. Sur-Kolay, A. Raghunathan, N.K. Jha, Systematic poisoning attacks on and defences for machine learning in healthcare, IEEE J. Biomed. Health Inform. 19 (6) (2014) 1893−1905.

[13] J.M. McGinnis, L. Stuckhardt, R. Saunders, M. Smith (Eds.), Best care at lower cost: the path to continuously learning health care in America, National Academies Press, Washington, DC, 2013.

[14] R. Kaushal, K.G. Shojania, D.W. Bates, Effects of computerized physician order entry and clinical decision support systems on medication safety: a systematic review, Arch. Intern. Med. 163 (12) (2003) 1409−1416.

[15] Y.Y. Han, J.A. Carcillo, S.T. Venkataraman, R.S. Clark, R.S. Watson, T.C. Nguyen, et al., Unexpected increased mortality after implementation of a commercially sold computerized physician order entry system, Pediatrics 116 (6) (2005) 1506−1512.

[16] E. Seto, K.J. Leonard, J.A. Cafazzo, J. Barnsley, C. Masino, H.J. Ross, Developing healthcare rule-based expert systems: case study of a heart failure telemonitoring system, Int. J. Med. Inform. 81 (8) (2012) 556−565.

[17] B. Otal, L. Alonso, C. Verikoukis, Highly reliable energy-saving MAC for wireless body sensor networks in healthcare systems, IEEE J. Sel. Areas Commun. 27 (4) (2009) 553−565.

[18] B.S. Raghavendra, D. Bera, A.S. Bopardikar, R. Narayanan, Cardiac arrhythmia detection using dynamic time warping of ECG beats in e-healthcare systems. IEEE international symposium on a world of wireless, mobile and multimedia networks, (2011), IEEE, pp. 1−6.

[19] T. Simonite, Google's AI eye doctor gets ready to go to work in India. Wired, Conde Nast, 20 November, (2018).

[20] Y. Wang, I.V. Tetko, M.A. Hall, E. Frank, A. Facius, K.F. Mayer, et al., Gene selection from microarray data for cancer classification—machine learning approach, Computat. Biol. Chem. 29 (1) (2005) 37−46.

[21] A. Esteva, B. Kuprel, R.A. Novoa, J. Ko, S.M. Swetter, H.M. Blau, et al., Dermatologist-level classification of skin cancer with deep neural networks, Nature 542 (7639) (2017) 115−118.

[22] T. Desautels, J. Calvert, J. Hoffman, M. Jay, Y. Kerem, L. Shieh, et al., Prediction of sepsis in the intensive care unit with minimal electronic health record data: a machine learning approach, JMIR Med. Inform. 4 (3) (2016) e28.

[23] R.F. Thompson, G. Valdes, C.D. Fuller, C.M. Carpenter, O. Morin, S. Aneja, et al., Artificial intelligence in radiation oncology: a specialty-wide disruptive transformation, Radiotherapy Oncol. 129 (3) (2018) 421−426.

[24] D. Karamshuk, A. Noulas, S. Scellato, V. Nicosia, C. Mascolo, Geo-spotting: mining online location-based services for optimal retail store placement. In *Proceedings of the 19th ACM SIGKDD international conference on Knowledge discovery and data mining*, (2013), *ACM*, pp. 793−801.

[25] Y. Kaneko, K. Yada, A deep learning approach for the prediction of retail store sales. In *2016 IEEE 16th international conference on data mining workshops (ICDMW)*, (2016), *IEEE*, pp. 531−537.

[26] M. Paolanti, M. Sturari, A. Mancini, P. Zingaretti, E. Frontoni, Mobile robot for retail surveying and inventory using visual and textual analysis of monocular pictures based on deep learning. In *2017 European conference on mobile robots (ECMR)*, (2017), *IEEE*, pp. 1−6.

[27] M. Lek, B. Anadarajah, N. Cerpa, R. Jamieson, Data mining prototype for detecting ecommerce fraud. *ECIS 2001 Proceedings*, (2001), 60.

[28] L.M. Minga, Y.Q. Feng, Y.J. Li, Dynamic pricing: ecommerce-oriented price setting algorithm. In *Proceedings of the 2003 international conference on machine learning and cybernetics, IEEE*, (2003), vol. 2, pp. 893−898.

[29] J.L. Berral, Í. Goiri, R. Nou, F. Julià, J. Guitart, R. Gavaldà, et al., Towards energy-aware scheduling in data centers using machine learning. In *Proceedings of the 1st international conference on energy-efficient computing and networking*, (2010), *ACM*, pp. 215-224.

[30] A. Tsanas, A. Xifara, Accurate quantitative estimation of energy performance of residential buildings using statistical machine learning tools, Energy Build. 49 (2012) 560−567.

[31] T. Hu, Y. Fei, QELAR: a machine-learning-based adaptive routing protocol for energy-efficient and lifetime-extended underwater sensor networks, IEEE Trans. Mob. Comput. 9 (6) (2010) 796−809.

[32] P. Baldi, P. Sadowski, D. Whiteson, Searching for exotic particles in high-energy physics with deep learning, Nat. Commun. 5 (2014) 4308.

[33] A. Radovic, M. Williams, D. Rousseau, M. Kagan, D. Bonacorsi, A. Himmel, et al., Machine learning at the energy and intensity frontiers of particle physics, Nature 560 (7716) (2018) 41–48.

[34] L. Monostori, A. Márkus, H. Van Brussel, E. Westkämpfer, Machine learning approaches to manufacturing, CIRP Ann. 45 (2) (1996) 675–712.

[35] T. Wuest, D. Weimer, C. Irgens, K.D. Thoben, Machine learning in manufacturing: advantages, challenges, and applications, Prod. & Manuf. Res. 4 (1) (2016) 23–45.

[36] D.T. Pham, A.A. Afify, Machine-learning techniques and their applications in manufacturing, Proc. Inst. Mech. Eng. Part B: J. Eng. Manufacture 219 (5) (2005) 395–412.

[37] N. Bhatla, K. Jyoti, An analysis of heart disease prediction using different data mining techniques, Int. J. Eng. 1 (8) (2012) 1–4.

[38] S. Dua, U.R. Acharya, P. Dua, Machine learning in healthcare informatics, Intelligent Systems Reference Library, Springer Nature, 2014, p. 56.

[39] J.T. Pollettini, S.R. Panico, J.C. Daneluzzi, R. Tinós, J.A. Baranauskas, A.A. Macedo, Using machine learning classifiers to assist healthcare-related decisions: classification of electronic patient records, J. Med. Syst. 36 (6) (2012) 3861–3874.

[40] S. Reddy, J. Fox, M.P. Purohit, Artificial intelligence-enabled healthcare delivery, J. R. Soc. Med. 112 (1) (2019) 22–28.

[41] R. Culkin, S.R. Das, Machine learning in finance: the case of deep learning for option pricing, J. Invest. Manag. 15 (4) (2017) 92–100.

[42] P. Treleaven, R.G. Brown, D. Yang, Blockchain technology in finance, Computer 50 (9) (2017) 14–17.

[43] G. Dimitrakopoulos, P. Demestichas, Intelligent transportation systems, IEEE Vehicular Technol. Mag. 5 (1) (2010) 77–84.

[44] J.M. Matías, J. Taboada, C. Ordóñez, P.G. Nieto, Machine learning techniques applied to the determination of road suitability for the transportation of dangerous substances, J. Hazard. Mater. 147 (1-2) (2007) 60–66.

[45] I. Habernal, T. Ptáček, J. Steinberger, Sentiment analysis in czech social media using supervised machine learning. In *Proceedings of the 4th workshop on computational approaches to subjectivity, sentiment and social media analysis*, (2013), pp. 65–74.

[46] D. Ruths, J. Pfeffer, Social media for large studies of behavior, Science 346 (6213) (2014) 1063–1064.

[47] S. Liu, S. Liu, W. Cai, S. Pujol, R. Kikinis, D. Feng, Early diagnosis of Alzheimer's disease with deep learning. In 2014 IEEE 11th international symposium on biomedical imaging (ISBI), IEEE, (2014), pp. 1015–1018.

[48] L.R. Trambaiolli, A.C. Lorena, F.J. Fraga, P.A. Kanda, R. Anghinah, R. Nitrini, Improving Alzheimer's disease diagnosis with machine learning techniques, Clin. EEG Neurosci. 42 (3) (2011) 160–165.

[49] L.N. Li, J.H. Ouyang, H.L. Chen, D.Y. Liu, A computer aided diagnosis system for thyroid disease using extreme learning machine, J. Med. Syst. 36 (5) (2012) 3327–3337.

[50] S. Ghumbre, C. Patil, A. Ghatol, Heart disease diagnosis using support vector machine. In international conference on computer science and information technology (ICCSIT'), Pattaya, (2011).

[51] J.S. Sartakhti, M.H. Zangooei, K. Mozafari, Hepatitis disease diagnosis using a novel hybrid method based on support vector machine and simulated annealing (SVM-SA), Comput. Methods Prog. Biomed 108 (2) (2012) 570−579.

[52] M.S. Bascil, F. Temurtas, A study on hepatitis disease diagnosis using multilayer neural network with Levenberg Marquardt training algorithm, J. Med. Syst. 35 (3) (2011) 433−436.

[53] R. Challen, J. Denny, M. Pitt, Artificial intelligence, bias and clinical safety, BMJ Qual. Saf. 28 (2019) 231−237.

[54] D. Pyle, C. San Jose, An executive's guide to machine learning, McKinsey Q. 3 (2015) 44−53.

[55] R. Parasuraman, D.H. Manzey, Complacency and bias in human use of automation: an attention integration, Hum. Factors 52 (3) (2010) 381−410.

[56] M. Chen, Y. Hao, K. Hwang, L. Wang, L. Wang, Disease prediction by machine learning over big data from healthcare communities, IEEE 5 (2017) 8869−8879.

CHAPTER

Machine learning for biomedical signal processing

3

Vandana Patel and Ankit K. Shah

Department of Instrumentation and Control Engineering, Lalbhai Dalpatbhai College of Engineering, Ahmedabad, India

3.1 Introduction

Machine learning (ML) is a tool which applies high computational based algorithms to get useful information and to optimize performance parameters from any given data set. There are many applications of ML-based techniques in various domains, such as video surveillance, speech recognition, image processing, news classification, weather predication, share market analysis, and medical services [1]. Recently, ML has been adopted widely in the area of biomedical applications by healthcare professionals to diagnose patients more accurately and/or instantaneously. ML can also be applied for the prediction of disease severity and planning the treatment or therapy accordingly.

Biomedical signals are primarily electrical signals which are generated due to the electrochemical activity of certain cells [2]. By measuring these signals, the health information of a person can be obtained. This information can be extracted through various types of biomedical signals, such as an electrocardiogram (ECG), electroencephalogram (EEG), electromyogram (EMG), electroretinogram (ERG), and electrooculogram (EOG). While capturing these signals using a suitable electrode, the useful signal is contaminated with various types of noise and unwanted artifacts. Biomedical signal processing deals with the analysis of these measured data to obtain useful information. In order to get useful information the filtering, feature extraction, and classification of the signal is required using suitable mathematical formulae, algorithms, or engineering tools. Traditionally biomedical signals were captured and computed by software, to support health care professionals with real-time data, thus helping to determine the state of a patient's health. The ML-based technique in signal processing improves data quality and provides accurate decision, and ultimately eliminates the traditional educated guesswork of healthcare professionals.

The death rate due to cardiac disorder has been drastically increasing over the past few decades. ECG is the most common medical tool used by cardiologists to monitor and diagnose the state of a heart condition. The heart has a special

Machine Learning and the Internet of Medical Things in Healthcare. DOI: https://doi.org/10.1016/B978-0-12-821229-5.00002-1
Copyright © 2021 Elsevier Inc. All rights reserved.

47

48 **CHAPTER 3** Machine learning for biomedical signal processing

electrical system which causes its atrium and ventricles to expand and contract. ECG is the measurement of the bioelectric activity of the heart by placing suitable electrodes on the patient's body. While capturing and transmitting the ECG signals, it gets corrupted due to various types of artifacts and unwanted interference, such as muscle noise (high frequency), baseline drift due to electrode movement, 50 Hz powerline interference, etc. To extract more precise and meaningful information from ECG, these artifacts have to be removed using suitable filters. The ML-based adaptive filters for preprocessing of the ECG signal provides better results compared to traditional analog and digital filters. The ML algorithms provide accurate and faster techniques for automatic ECG feature extraction, decision-making. and classification of heart abnormalities [3].

In this chapter, various ML algorithms for processing biomedical signals, particularly ECG, are discussed. The chapter is organized as follows. In Section 3.2, the morphology and measurement of the ECG signal is briefly discussed. Various stages for processing the ECG signal are also discussed along with the ECG block diagram. The application of an adaptive filter for preprocessing of the raw ECG signal along with various adaptive algorithms is discussed in Section 3.3. In the same section, the responses of different adaptive filter algorithms for the ECG signal are shown. In Section 3.4, different ML-based algorithms proposed for the feature extraction and classification are discussed. Section 3.5 provides a conclusion on various aspects of ML-based techniques for biomedical signal processing.

3.2 Reviews of ECG signal

The ECG is a measurement of the amplitude and time intervals of the heart-generated bioelectrical activity. These bioelectrical signals are generated due to the depolarization and repolarization of cardiac cells present in atriums and ventricles. The bioelectrical activity of the heart is measured by placing suitable electrode leads on various parts of the body [4].

The standard ECG signal is recorded using Bipolar Standard Limb Leads (three electrodes), Unipolar Augmented Limb Leads (three electrodes), and Precordial Leads (six electrodes). Electrodes are placed on three limbs namely right arm (RA), left arm (LA), and left leg (LL) to form a triangle know as Einthoven's Equilateral Triangle [4], as in Fig. 3.1.

The standard limb leads are obtained by taking the potential difference between two limb leads and are known as lead-I, lead-II, and lead-III, as mentioned in Table 3.1. As per Einthoven's law the vector sum of lead-I and lead-III is equal to lead-II.

The augmented limb leads are unipolar and obtained by placing a null point as zero reference potential, as shown in Fig. 3.2. The potential difference between the null point and each limb lead are known as aVR, aVL, and aVF, as mentioned in Table 3.2. The augmented limb leads give the additional information by recording the potential difference across the heart in three different directions [4].

3.2 Reviews of ECG signal

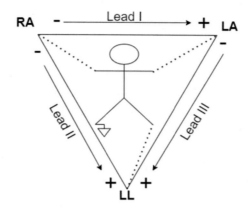

FIGURE 3.1
Einthoven's equilateral triangle based electrode placement.

Table 3.1 ECG Lead electrodes.

Lead	Electrode placement with polarity	
I	RA-Negative	LA-Positive
II	RA-Negative	LL-Positive
III	LA-Negative	LL-Positive

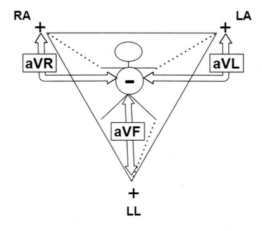

FIGURE 3.2
ECG augmented limb leads placement.

Table 3.2 ECG augmented limb leads electrodes.

Lead	Electrode placement with polarity where negative electrode is common	
aVR	RA	Positive
aVL	LA	Positive
aVF	LL	Positive

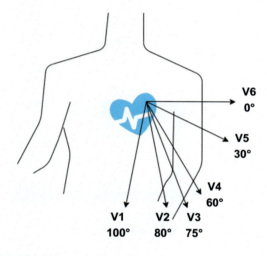

FIGURE 3.3

Precordial leads placement on chest.

The precordial leads are also known as chest leads and labeled V1–V6. The abovementioned limb leads measure electrical activity in up–down and right–left directions, while precordial leads measure electrical activity in back–front and right–left directions, as shown in Fig. 3.3.

Various dataset of ECG signal is available in the data base given at Massachusetts Institute of Technology–Beth Israel Hospital (MIT-BIH) [5].

One cardiac (depolarization–repolarization) cycle of an ECG waveform consists of PQRST(U) waves, as shown in Fig. 3.4. The normal ECG signal is analyzed by observing a P wave, QRS complex and T wave, PR interval, QRS complex (also called QRS duration) and QT interval, ST segment, and PR segment, as shown in Fig. 3.4.

The ECG signal is captured from electrodes often contaminated with various types of interference signals from other sources. The signal processing of ECG data is required in order to extract useful information from the ECG signal. The various stages of ECG signal processing are shown in Fig. 3.5.

In the subsequent sections, different ML-based techniques that are used for the ECG signal preprocessing and classification are discussed.

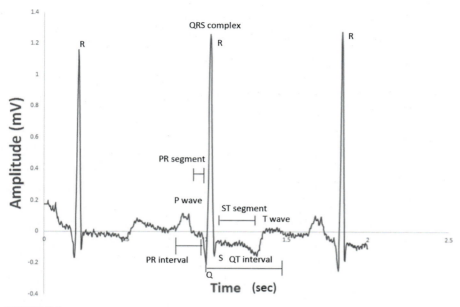

FIGURE 3.4

Normal ECG signal.

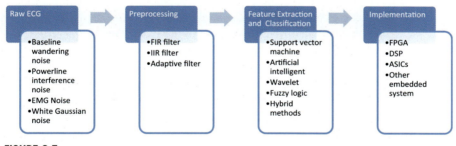

FIGURE 3.5

Block diagram of ECG signal processing.

3.3 Preprocessing of ECG signal using ML based techniques

Preprocessing of the ECG signal is essential to remove the undesired interfering elements and to enhance the desired feature from the biomedical signal. The ECG signal has a frequency range between 0.5 and 100 Hz with the major information related to electrical activity of heart is focused between 5 and 35 Hz, and the

52 **CHAPTER 3** Machine learning for biomedical signal processing

amplitude is around $1-2$ mV [6,7], which can be considered to be a weak physiological signal with low frequency. The ECG signal is usually recorded by placing suitable electrodes on the surface of the patient's body after applying electrolyte gel. The movement of either electrode or patient, improper contact between the electrode and the surface of the skin, or interference of other bioelectrical signals may lead to the inclusion of noises in the ECG signal. During transmission and data acquisition of the ECG signal, it may affected by high-frequency white Gaussian noise due to an improper channel, and 50/60-Hz powerline noise due to electromagnetic field interference.

With respect to the entire interference signal, the interference signals in the ECG may be classified into four types of artifact based on the frequency band.

- White Gaussian noise
  This high-frequency noise corrupts the ECG signal, primarily during wireless transmission of the ECG signal.
- EMG noise
  During ECG recording, bioelectrical signals are generated from other cells, such as skeletal muscles which are captured by ECG electrodes, resulting in the contamination of the ECG signal. Removal of such EMG noise in the ECG signal is a challenging task for filtering as the spectral frequency of muscle activity (EMG signal) coincides with the PQRST complex present in the ECG signal.
- Baseline wandering noise
  This is a very low frequency, slowly varying artifact generated due to changes in skin electrode impedance, movement in skin—electrode contact, respiration, etc. Baseline wandering noise in the ECG signal deviates the entire signal from its normal baseline.
- Powerline interference (PLI) noise
  PLI noise is identified as 50/60-Hz sinusoidal powerline interference with its harmonics generated due to the electromagnetic field.

Considering the need to extract the desired features from the ECG signal, the preprocessing technique is mainly focused toward removal of unwanted noise components of the ECG signal. These unwanted noisy components can be removed by selecting the appropriate filter or filter bank with fixed coefficients. The time-invariant filters have a fixed structure along with fixed internal parameters. The ECG signals are time-varying signals with a nonstationary environment; while dealing with such signals whose properties vary with time, fixed filter algorithms may not process such signals significantly. The optimal solution is to apply an ML-based adaptive algorithm for filtering the ECG signal in such a nonstationary environment.

An adaptive filter can automatically update its coefficients/parameters with changes in the environment, in order to get optimal performance parameters. The comparative study of performance criteria for the elimination of PLI noise from the ECG signal by applying adaptive and nonadaptive filters is done

3.3 Preprocessing of ECG signal using ML based techniques

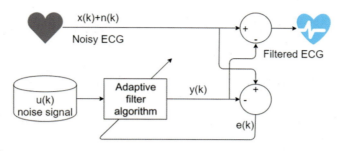

FIGURE 3.6

Adaptive filter configuration for signal enhancement.

by Hamilton [8]. An adaptive filter will improve the performance with less complexity and better noise rejection from the ECG signal. Thakor and Zhu [9] developed various structures of an adaptive filter to remove noise present in the ECG signal. They used an adaptive recurrent structure to detect cardiac arrhythmias. An adaptive filter configuration is shown in Fig. 3.6 for processing the ECG signal.

In Fig. 3.6, two input signals are given. One is the reference signal consisting of a desired ECG signal x(k) and a noise n(k) which is denoted as unwanted signals present in the ECG. The second input signal is a noise signal u(k) which is correlated with the interference signal n(k), but uncorrelated with the desired ECG signal x(k). The signal u(k) is applied to the adaptive filter. The error signal e(k) is given by

$$e(k) = x(k) + n(k) - y(k) \quad (3.1)$$

In Eq. (3.1), y(k) is the output of the adaptive filter defined as

$$y(k) = \sum_{l=0}^{N} w_l n(k-l) \quad (3.2)$$

where w_l are weight coefficients. The resulting mean square error (MSE) is then given by

$$E[e^2(k)] = E[x^2(k)] + E\{[n(k) - y(k)]^2\} \quad (3.3)$$

Based on the error signal e(k), the adaptive filter updates its coefficients w_l. As the new value of the input signals is received, the adaptive filter updates its coefficients in order to minimize the MSE as given in Eq. (3.3).

The learning rule adapted for modification in the coefficients is based on minimization of the error signal e(k) = d(k) − y(k). There are also various other algorithms proposed for elimination of the different types of noise discussed above based on their characteristics. The most widely used adaptive filtering algorithms for the removal of noise present in ECG signal are discussed below with their weight update equations, as given by Qureshi et al. [7].

54 **CHAPTER 3** Machine learning for biomedical signal processing

3.3.1 **Least mean square (LMS)**

The LMS algorithm adjusts the filter parameters in order to minimize the mean squares error between the filter output signal and the expectations output signals. LMS is based on a steepest descent algorithm. The updated filter coefficient for LMS algorithm is given by

$$w(k + 1) = w(k) + 2\mu * e(k) * x(k) \tag{3.4}$$

In the Eq. (3.4), μ is the step size. For the analysis of the adaptive filter algorithm the ECG data from the MIT-BIH database is used. This database is freely available for researchers. The ECG signal is taken directly from the database and then contaminated with 50 Hz PLI and baseline wandering noise obtained from the MIT-BIH database. The simulation is performed in MATLAB to analyze the performance of the LMS adaptive filter algorithms. Here, contaminated ECG with PLI and baseline wandering noise is given at the input node of the adaptive filter and the known ECG noise is applied at the desired node. The performance of the LMS-based adaptive filter with mean square error is shown in Fig. 3.7.

3.3.2 **Normalized least mean square (NLMS)**

To overcome the issue of appropriate selection of step size, the NLMS algorithm is implemented. Here step size selection is adaptive which leads to fast convergence and stability. The updated filter coefficient for the NLMS algorithm is defined as

$$w(k + 1) = w(k) + \frac{2\mu}{x(n)x^T(n)} e(k) * x(k) \tag{3.5}$$

where $\frac{2\mu}{x(n)x^T(n)}$ is the normalized step size. The NLMS adaptive algorithm is applied to the noisy ECG signal with PLI and baseline wandering noise. The performance of the NLMS-based adaptive filter with mean square error is shown in Fig. 3.8.

3.3.3 **Delayed error normalized LMS (DENLMS) algorithm**

The DENLMS adaptive algorithm is preferred when high computational speed is required with less complexity. The updated filter coefficient for the DENLMS algorithm is given by

$$w(k + 1) = w(k) + \mu e(k - nD) * x(k - nD) \tag{3.6}$$

where μ is the step size and nD denotes the number of delays element. In [10], the Gaussian noise is eliminated from the ECG signal using a DENLMS-based adaptive filter algorithm. The performance enhancement and reduction in the power consumption are achieved using the pipeline structure of the DENLMS algorithm.

3.3 Preprocessing of ECG signal using ML based techniques

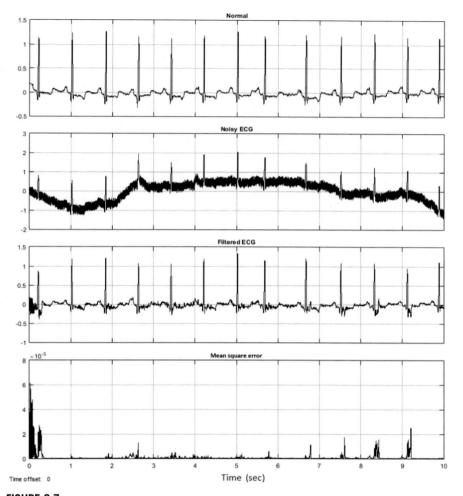

FIGURE 3.7

LMS adaptive filter result.

3.3.4 Sign data least mean square (SDLMS)

SDLMS is obtained by substituting the input x(k) with sgn(k) LMS algorithm which makes SDLMS less complex than LMS. Initially, it converges slowly, but later on it speeds up as the MSE value drops. Generally, this algorithm is preferred where high-speed computation is required, such as in a biotelemetry application. The updated filter coefficient for SDLMS algorithm is given by

$$w(k+1) = w(k) + 2\mu * e(k) * sgn(k) \quad (3.7)$$

56 CHAPTER 3 Machine learning for biomedical signal processing

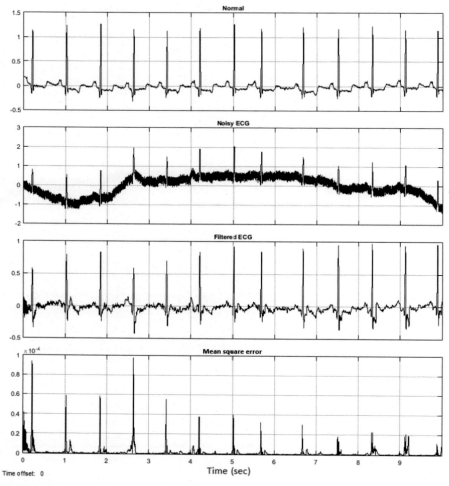

FIGURE 3.8

NLMS adaptive filter result.

where

$$\text{sgn}(k) = \begin{cases} 1, & x < 0 \\ 0, & x = 0 \\ -1, & x > 0 \end{cases}$$

The SDLMS adaptive algorithm is applied to the noisy ECG signal with PLI and baseline wandering noise. The performance of the SDLMS-based adaptive filter with mean square error is shown in Fig. 3.9.

3.3 Preprocessing of ECG signal using ML based techniques

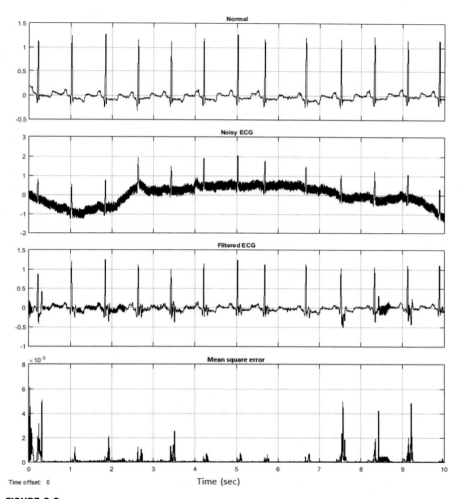

FIGURE 3.9

SDLMS adaptive filter result.

3.3.5 Log least mean square (LLMS)

Here, quantization of the error signal is done instead of the input signal. The updated filter coefficient for LLMS algorithm is given by

$$w(k+1) = w(k) + 2\mu * Qe(k) * x(k) \tag{3.8}$$

where Qe(k) is the quantization function and the logarithmic function Q is given by,

$$Q(z) = 2^{\log_2(z)} \text{sgn}(z) \tag{3.9}$$

58 CHAPTER 3 Machine learning for biomedical signal processing

The elimination of several artifacts which degrade the quality of signal and lead to loss of information is the key role of adaptive filters. Here, the knowledge and characteristics of the noise signal is required a priori for the best results. The objective of the adaptive algorithm is to minimize the MSE and to improve the ECG signal for the extraction of useful information. Compared to LMS, NLMS provides a better MSE performance, but at the cost of more calculation and storage. SDLMS algorithm is preferred for faster results due to its high-speed computation.

Various adaptive filter algorithms have been compared for the removal of stationary and nonstationary noises from ECG [11]. The performance of algorithms are verified by means of structure complexity, computational complexity, and signal-to-noise ratio. The EMG noise is removed from the ECG signal using the recursive least squares-based adaptive algorithm in [12]. The performance of the proposed algorithm is checked using parameters, such as order of the filter, signal-to-noise ratio, forgetting factor, etc. In [13], the motion artifacts are removed from the ECG signal using a cosine transform LMS-based adaptive filter algorithm. The proposed algorithm is able to removed high-amplitude noise more efficiently compared to classical LMS adaptive algorithms. However, one can also design multistage adaptive filters for ECG filtering, as the ECG signal is contaminated with different types of artifacts with distinct frequency spectra.

3.4 Feature extraction and classification of ECG signal using ML-based techniques

There are many challenges in extracting features from the ECG signal and classification of the signal. Major challenges in ECG signal classification are due to varying features of ECG signals, the lack of standardized features, variability in ECG patterns of diseases/normal person to person [1], etc. Many researchers are working on ML-based classifiers which are capable of classifying arrhythmia. However, the ML-based classifier result has the main disadvantage of its dependency on heuristic knowledge. In ML-based algorithms the learning process is mainly three types: supervised learning, unsupervised learning, and reinforcement learning. The supervised learning algorithm uses labeled data set by experience personnel. This type of learning requires expertise and is time consuming. The unsupervised learning finds novel diseases from the pattern of the data without any human feedback. However, it may produce results with bias as they are not validated with another group of data. The combined learning of supervised and unsupervised is called reinforcement learning. However, to get better performance the reinforcement learning requires more time and training data.

The effectiveness of the classification algorithm is generally validated with statistical terms like sensitivity (Se), positive prediction (+P), specificity (Sp), and accuracy (Ac) [14]. The abovemention statistical measures are defined as:

$$Se = \frac{TP}{TP + FN} \tag{3.10}$$

$$+P = \frac{TP}{TP + FP} \tag{3.11}$$

$$Sp = \frac{TN}{TN + FP} \tag{3.12}$$

$$Ac = \frac{TP + TN}{TP + TN + FN + FP} \tag{3.13}$$

where TP is true positives, TN is true negatives, FP is false positives, and FN is false negatives. If the doctor's decision is matched with arrhythmia detection of the classifier then the TP decision is generated. If the doctor's decision as well as arrhythmia detection of the classifier is negative, then the TN decision is generated. The FP decision occurs if the classifier labels an arrhythmia to a healthy patient. The FN decision occurs if the classifier labels a healthy to an arrhythmia patient. A good classification algorithm minimizes FN and FP decisions. Another statistical parameter F1 score is used as a function of Se and $+P$. The F1 score is defined as

$$F1 = \frac{2(Se^* + P)}{(Se + +P)} \tag{3.14}$$

In the training phase, the F1 score is used as an optimizing parameter for the classifier. Various methods for the ECG signals classifiers and feature extraction have been proposed, such as Pan and Tompkins method, neural network, wavelet transform, support vector machine, self-organizing map, fuzzy logic, hybrid methods, etc. Here, various ML-based algorithms are discussed as below.

3.4.1 Artificial neural network (ANN)

ANN is a biological activity-based network which is suitable for classifying biomedical signals. There are several advantages to using ANN-based ML algorithms, such as their self-adaptive nature, they are fast, accurate, and handle nonlinearities. ANN is also scalable and robust against noisy data. ANN is applicable to the ECG signal due to its main advantages being (1) the ECG signal is nonlinear by nature and ANN handles nonlinearities using different activation functions; (2) ECG signals have lower frequencies and ANN have an adaptive model for lower frequencies; and (3) the ECG signal has time varying and nonlinear noises and the ANN removes these noises.

However, the ANN has several issues regarding its implementation like (1) ANN requires a sufficient set of data for training and validation purposes; (2) ANN may not provide a global optimal solution for the 12-lead ECG classification; and (3) the classification results using ANN depend on the selection of number of neurons.

There have been many algorithms proposed based on ANN for the ECG signal classification. In [15], a block-based ANN is proposed for the ECG signal classification. A robust and generic ANN-based algorithm for classifying the ECG signal has been proposed in [16]. The Recurrent Neural Network (RNN) is combined with an Eigen vector-based method in [17]. In these, the Physiobank database is analyzed for classifying four types of ECG beats. The two classifiers-based RNN is proposed in [18] for the detection of heartbeats.

Recently, deep neural networks (DNN)-based classification has been used by many researchers. In [19], DNN-based active classification of ECG signals was proposed using learning and interaction phases. A deep convolutional neural network (CNN)-based nine-layer model was proposed in [20] for identifying five categories of heartbeats present in ECG signals. A DNN model architecture was proposed in [21] using a CNN and recurrent network. The device's independent and interpatient ECG arrhythmia classification method was developed.

A 1D-convolutional ANN model was proposed in [22] for the cardiac arrhythmia detection of ECG signals. Deep CNN-based arrhythmia detection technique was used in [23]. A CNN algorithm is proposed in [24] for detection of myocardial infarction (MI) from the ECG signals. They achieved 93.53% accuracy with noise present in the signal. RNN is used in [25] to find the beat from the ECG. The global and updatable ECG beat classification is obtained from the RNN which is based on morphological and temporal information.

3.4.2 Fuzzy logic (FL)

The FL-based ML is an approach used for ECG signal classification and biomedical systems. The FL can include the linguistic approximation which represents knowledge in a more useful and meaningful way. This specialty of the FL is important in ECG signal classification as it involves decision-making based on knowledge. The FL uses smooth function variables with different membership functions. The FL can also handle nonlinearities by means of the defining rules in expert systems.

A fuzzy rule-based set support decision system was proposed from the ECG records by [26]. They developed a decision system for the disease identification. An adaptive fuzzy ECG classifier-based statistical learning algorithm was proposed with fixed parameters in [27]. The proposed classifier had an average correct rate of 88.2%. The ECG signal also has information regarding coronary artery diseases. In [28], a fuzzy expert system was developed for the prediction of risk status of coronary artery diseases. They emphasized a tree structure for the search rule organization from a large set of data to get a precise prediction of diseases. The fuzzy-based inference system was proposed for denoising the ECG signal in [29]. The MSE and signal-to-noise ratio is improved using the fuzzy inference system.

The FL is also used for the clustering of the data points known as fuzzy c-means (FCM) clustering. The fuzzy-based FCM clustering method was developed and improved by Bezdek in 1974 and it is used for data classification by researchers in various applications [30]. The FCM-based clustering is also useful for classifying the ECG signal. FCM was used for the classification of 10 different types of arrhythmias from the ECG signal in [31]. The records of 92 patients were used to compare the results. The FCM clustering reduced the number of segments which enhanced the speed of the algorithm by grouping similar segments of the training data. The type-2 FCM clustering algorithm is also used for arrhythmia classification from ECG in [32]. They increased the efficiency in the training and testing phase using FCM. By selecting the best arrhythmia they formed a new training data set using type-2 FCM. The 99% accuracy rate is achieved by the use of a clustering-based technique. In [33], FCM is used to classify the heartbeat cases from the ECG signals. The proposed method was able to classify abnormal and normal heartbeats with an accuracy of 93.57%. The FCM spectral model-based left ventricular heart failure algorithm was developed from ECG and cardiac color ultrasound in [34].

3.4.3 **Wavelet transforms**

The wavelet transform was used by Morlet in the early 1980s. The wavelet transform can synthesize one- and multidimensional data which provide an alternate solution to the Fourier transform. The wavelet transform has various applications in the field of data analysis and classification, such as communication, audio—video signal processing, image processing, and biomedical signal processing. The wavelet transform method is used for the continuous time signal and discrete time signal. The wavelet transform is applicable for the ECG signals as they are non-stationary signals. The wavelet transform is found by convolving low-frequency filters and high-frequency filters to the signals and then further downsampling. In this way, wavelet coefficients are able to measure small variations around the data point. This feature of the wavelet is useful to find spikes or discontinuities in the ECG signals.

The hidden difficulties and challenges present in the ECG signals are poorly defined by the continuous time domain and frequency domain compared to wavelet transform [35]. The QRS complex wave was detected using the wavelet transforms in [36]. The P and T waves were also detected from the ECG signal with noise. They achieved a QRS complex detection rate above 99.8% with drift and noise present in the ECG signal. In [37], the QRS complex detector was developed using the dyadic wavelet transform from the ECG signals. The proposed algorithm provided superior results compared to other statistical algorithms. In [38], the wavelet transform is used to reduce the dimensions of the ECG signal and then fed to neural network. By using the proposed wavelet method, the values obtained for heartbeat classification parameter Ac, Sp, Se and + P are 99.28%, 99.83%, 99.97% and 99.21%, respectively. The RR intervals were extracted from the ECG signal using the discrete wavelet transform in [39]. The proposed structure reduced the noninformative functions and helped to reduce the data length of

62 CHAPTER 3 Machine learning for biomedical signal processing

the classifier. Various critical issues like scalability and stability overtime have been addressed.

In [40], the fast Fourier transform (FFT) and discrete wavelet transform were used to get a frequency component from the ECG signals in order to classify the signal. A classification of arrhythmia using the morphological features of the discrete wavelet transform was proposed in [41]. The information of RR intervals is defined as a dynamic feature of the ECG signal. The algorithm was tested over MIT-BIH arrhythmia database and it was proved that it improved the accuracy of classification. The QRS complex detection technique was developed using the tunable Q wavelet transform in [42]. The optimally tuned parameters and subband improve the decomposition of the ECG signal. The proposed method has 99.88% accuracy, 99.89% sensitivity, and 99.83% positive productivity for the ECG signals given in MIT-BIH arrhythmia database.

3.4.4 Hybrid approach

The ML-based algorithm is used with a combination of statistical algorithms and other ML algorithms. A combination of statistical algorithms and ML algorithms was called a hybrid approach for the ECG signal.

The fuzzy neural network for beat recognition and classification of ECG was developed in [43]. The FCM and ANN have been applied to recognize different types of beats from the ECG waveform and results shows good efficiency. In [44], wavelet transform is used for feature extraction and a fuzzy neural-based hybrid model was used for the classification of the ECG beats. The integrated independent component analysis (ICA) along with ANN-based beat classification from the ECG signals was proposed in [45]. Here, the ECG signal is decomposed into weighted sum of statistically mutual independent basic components using ICA method. The probabilistic neural network (PNN) with back-propagation neural network (BPNN) is applied to classify the data. The proposed algorithm gives a classification accuracy of over 98%. Fuzzy neural networks (FNNs) have been proposed in [46] to detect premature ventricular contractions (PVCs) from the ECG signal. The ANN is used with weighted fuzzy membership functions to train the ECG signal from the MIT-BIH PVC database.

The particle swarm optimization (PSO) and radial basis function neural network (RBFNN)-based ECG beat classification method was developed in [47]. Six different types of beats classification have been obtained with a smaller size of network. The fusion approach of SVM, wavelet, and ANN is used in [48]. First they obtained the ECG signal using robust discrete wavelet transform. Next they proposed a ML algorithm to find arrhythmias and achieved 98.06% accuracy. The principle component analysis (PCA)-based classification algorithm combined with the DNN to find the arrhythmia classification in the ECG signal in [49]. The PCA was used to reduce the unbalance data structure present in the signal. They achieved accuracy of 96.7% with the hybrid approach.

3.5 Discussions and conclusions

In this chapter, various machine learning-based algorithms and their applications in biomedical signal processing have been discussed. First, we briefly introduced the importance of ML-based techniques for biomedical signal processing. In the rest of the chapter, we have mainly focused on ECG signal processing. Various learning approaches of LMS-based adaptive algorithms have been discussed for the preprocessing of the ECG signal. The performance of different adaptive filter algorithms is shown by means of a simulation in MATLAB using the MIT-BIH database. Various ML-based techniques are briefly explained for the ECG signal feature extraction and classification. Different applications and literature based on ANN, FL, wavelet, and hybrid approaches for the ECG signal are discussed for cardiac arrhythmia. The theoretical concepts of ML applied for the real-time applications of biomedical signal processing are demonstrated. The ML-based technique can be implemented to design a low-power, high-speed, low-cost, and accurate device, using the appropriate embedded platform for real-time diagnosis applications in the biomedical field.

References

[1] K.R. Foster, R. Koprowski, J.D. Skufca, Machine learning, medical diagnosis, and biomedical engineering research - commentary, Biomed. Eng. Online 13 (1) (2014).

[2] E. Ozpolat, B. Karakaya, T. Kaya, A. Gulten, FPGA-based digital filter design for biomedical signal, *Perspect. Technol. Methods MEMS Des. MEMSTECH 2016 - Proc. 12th Int. Conf.*, pp. 70–73, (2016).

[3] C.K. Roopa, B.S. Harish, A survey on various machine learning approaches for ECG analysis, Int. J. Comput. Appl. 163 (9) (2017) 25–33.

[4] John G. Webster, Medical instrumentation application and design, 4th ed., Wiley, 2010.

[5] A.L. Goldberger, L.A.N. Amaral, L. Glass, J.M. Hausdorff, PCh. Ivanov, R.G. Mark, et al., PhysioBank, PhysioToolkit, and PhysioNet: Components of a new research resource for complex physiologic signals. (2003).

[6] J. Zhang, B. Li, K. Xiang, X. Shi, Method of diagnosing heart disease based on deep learning ECG signal. (2019).

[7] R. Qureshi, M. Uzair, K. Khurshid, Multistage adaptive filter for ECG signal processing, *Proc. 2017 Int. Conf. Commun. Comput. Digit. Syst. C-CODE 2017*, (2017) pp. 363–368.

[8] P.S. Halmiton, A comparison of adaptive and nonadaptive filters for reduction of power line interference in the ECG, IEEE Trans. Biomed. Eng. 43 (1) (1996) 105–109.

[9] N.V. Thakor, Y.-S. Zhu, Applications of adaptive filtering to ECG analysis: noise cancellation and arrhythmia detection, IEEE Trans. Biomed. Eng. 38 (8) (1991) 785–794.

[10] C. Venkatesan, P. Karthigaikumar, R. Varatharajan, FPGA implementation of modified error normalized LMS adaptive filter for ECG noise removal, Clust. Comput. (2018) 1–9.

[11] M.Z.U. Rahman, R.A. Shaik, D.V.R.K. Reddy, Efficient and simplified adaptive noise cancelers for ecg sensor based remote health monitoring, IEEE Sens. J. 12 (3) (2012) 566−573.

[12] G. Lu, et al., Removing ECG noise from surface EMG signals using adaptive filtering, Neurosci. Lett. 462 (1) (2009) 14−19.

[13] F. Xiong, D. Chen, Z. Chen, S. Dai, Cancellation of motion artifacts in ambulatory ECG signals using TD-LMS adaptive filtering techniques, J. Vis. Commun. Image Represent. 58 (2019) 606−618.

[14] M. Alfaras, M.C. Soriano, S. Ortín, A fast machine learning model for ECG-based heartbeat classification and arrhythmia detection, Front. Phys. 7 (2019).

[15] W. Jiang, S.G. Kong, Block-based neural networks for personalized ECG signal classification, IEEE Trans. Neural Netw. 18 (6) (2007) 1750−1761.

[16] T. Ince, S. Kiranyaz, M. Gabbou, A generic and robust system for automated patient-specific classification of ECG signals, IEEE Trans. Biomed. Eng. 56 (5) (2009) 1415−1426.

[17] E.D. Übeyli, Combining recurrent neural networks with eigenvector methods for classificationof ECG beats, Digit. Signal. Process. 19 (2) (2009) 320−329.

[18] T. Teijeiro, P. Félix, J. Presedo, D. Castro, Heartbeat classification using abstract features from the abductive interpretation of the ECG, IEEE J. Biomed. Heal. Inform. 22 (2) (2018) 409−420.

[19] M.M. Al Rahhal, Y. Bazi, H. AlHichri, N. Alajlan, F. Melgani, R.R. Yager, Deep learning approach for active classification of electrocardiogram signals, Inf. Sci. 345 (2016) 340−354.

[20] U.R. Acharya, et al., A deep convolutional neural network model to classify heartbeats, Comput. Biol. Med. 89 (2017) 389−396.

[21] L. Guo, G. Sim, B. Matuszewski, Inter-patient ECG classification with convolutional and recurrent neural networks, Biocybern. Biomed. Eng. 39 (3) (2019) 868−879.

[22] S. Kiranyaz, T. Ince, M. Gabbouj, Real-time patient-specific ECG classification by 1-D convolutional neural networks, IEEE Trans. Biomed. Eng. 63 (3) (2016) 664−675.

[23] Ö. Yıldırım, P. Pławiak, R.S. Tan, U.R. Acharya, Arrhythmia detection using deep convolutional neural network with long duration ECG signals, Comput. Biol. Med. 102 (2018) 411−420.

[24] U.R. Acharya, H. Fujita, S.L. Oh, Y. Hagiwara, J.H. Tan, M. Adam, Application of deep convolutional neural network for automated detection of myocardial infarction using ECG signals, Inf. Sci. (Ny). 415−416 (2017) 190−198.

[25] G. Wang, et al., A global and updatable ECG beat classification system based on recurrent neural networks and active learning, Information Science, Elsevier Inc, 2018.

[26] S. Mitra, M. Mitra, B.B. Chaudhuri, An approach to a rough set based disease inference engine for ECG classificationLNAI 4259 - Proc. 5th Int. Conf. Rough. Sets Curr. Trends Comput. (2006).

[27] W.K. Lei, B.N. Li, M.C. Dong, M.I. Vai, AFC-ECG: an adaptive fuzzy ECG classifier, Soft Comput. Ind. Appl. (2007) 189−199.

[28] D. Pal, K.M. Mandana, S. Pal, D. Sarkar, C. Chakraborty, Fuzzy expert system approach for coronary artery disease screening using clinical parameters, Knowl. Syst. 36 (2012) 162−174.

References

[29] S. Goel, P. Tomar, G. Kaur, A fuzzy based approach for denoising of ECG signal using wavelet transform, Int. J. Bio-Sci. Bio-Technol. 8 (2) (2016) 143−156.

[30] A.K. Shah, D.M. Adhyaru, Parameter identification of PWARX models using fuzzy distance weighted least squares method, Appl. Soft Comput. 25 (2014) 174−183.

[31] Y. Özbay, R. Ceylan, B. Karlik, A fuzzy clustering neural network architecture for classification of ECG arrhythmias, Comput. Biol. Med. 36 (4) (2006) 376−388.

[32] R. Ceylan, Y. Özbay, B. Karlik, A novel approach for classification of ECG arrhythmias: type-2 fuzzy clustering neural network, Expert. Syst. Appl. 36 (3) (2009) 6721−6726. PART 2.

[33] Y.C. Yeh, W.J. Wang, C.W. Chiou, A novel fuzzy c-means method for classifying heartbeat cases from ECG signals, Meas. J. Int. Meas. Confed. 43 (10) (2010) 1542−1555.

[34] J. Dongdong, N. Arunkumar, Z. Wenyu, L. Beibei, Z. Xinlei, Z. Guangjian, Semantic clustering fuzzy c means spectral model based comparative analysis of cardiac color ultrasound and electrocardiogram in patients with left ventricular heart failure and cardiomyopathy, Futur. Gener. Comput. Syst. 92 (2019) 324−328.

[35] P.S. Addison, Wavelet transforms and the ECG: a review, Physiol. Meas. 26 (5) (2005) R155−R199.

[36] C. Li, C. Zheng, C. Tai, Detection of ECG characteristic points using wavelet transforms, IEEE Trans. Biomed. Eng. (1995).

[37] S. Kadambe, R. Murray, G. Faye Boudreaux-Bartels, Wavelet transform-based QRS complex detector, IEEE Trans. Biomed. Eng. (1999).

[38] R.J. Martis, U.R. Acharya, L.C. Min, ECG beat classification using PCA, LDA, ICA and discrete wavelet transform, Biomed. Signal. Process. Control. 8 (5) (2013) 437−448.

[39] M.M. Tantawi, K. Revett, A.B. Salem, M.F. Tolba, A wavelet feature extraction method for electrocardiogram (ECG)-based biometric recognition, Signal, Image Video Process. 9 (6) (2015) 1271−1280.

[40] Universiti Malaysia Perlis. School of Mechatronic Engineering, IEEE Malaysia Section, IEEE Engineering in Medicine and Biology Society. Malaysia Chapter, and Institute of Electrical and Electronics Engineers, Proceedings, 2015 2nd International Conference on Biomedical Engineering (ICoBE 2015) : 30−31 (2015), Penang, Malaysia.

[41] S.M. Anwar, M. Gul, M. Majid, M. Alnowami, Arrhythmia classification of ECG signals using hybrid features, Comput. Math. Methods Med. 2018 (2018).

[42] A. Sharma, S. Patidar, A. Upadhyay, U. Rajendra Acharya, Accurate tunable-Q wavelet transform based method for QRS complex detection, Comput. Electr. Eng. 75 (2019) 101−111.

[43] S. Osowski, T.H. Linh, ECG beat recognition using fuzzy hybrid neural network, IEEE Trans. Biomed. Eng. (2001).

[44] M. Engin, ECG beat classification using neuro-fuzzy network, Pattern Recognit. Lett. 25 (15) (2004) 1715−1722.

[45] S.N. Yu, K.T. Chou, Integration of independent component analysis and neural networks for ECG beat classification, Expert. Syst. Appl. 34 (4) (2008) 2841−2846.

[46] J.S. Lim, Finding features for real-time premature ventricular contraction detection using a fuzzy neural network system, IEEE Trans. Neural Netw. 20 (3) (2009) 522−527.

[47] M. Korürek, B. Doğan, ECG beat classification using particle swarm optimization and radial basis function neural network, Expert. Syst. Appl. 37 (12) (2010) 7563−7569.

[48] M.R. Homaeinezhad, S.A. Atyabi, E. Tavakkoli, H.N. Toosi, A. Ghaffari, R. Ebrahimpour, ECG arrhythmia recognition via a neuro-SVM-KNN hybrid classifier with virtual QRS image-based geometrical features, Expert. Syst. Appl. 39 (2) (2012) 2047–2058.

[49] S. Nurmaini, P.R. Umi, M. Naufal, A. Gani, Cardiac arrhythmias classification using deep neural networks and principle component analysis algorithm, Int. J. Adv. Soft Comput. Appl. (2018).

CHAPTER

Artificial itelligence in medicine

4

Arun Kumar Singh[1], Ashish Tripathi[1], Krishna Kant Singh[2], Pushpa Choudhary[1] and Prem Chand Vashist[1]

[1]*Department of Information Technology, G. L. Bajaj Institute of Technology and Management, Greater Noida, India*
[2]*Faculty of Engineering & Technology, Jain (Deemed-to-be University), Bengaluru, India*

4.1 Introduction

In modern life, disease is an incident that happens in the lifetime of people. Health is a foundation of a good life, but it is difficult to manage health due to unwanted threats; here threats are the types of diseases. The concept of disease is important because disease requires accurate concentration and care as the disease is associated with suffering. In many countries there are rights to treatment against diseases, so the significance of health insurers and policy makers is high. The disease can be defined by mental or biological phenomena described in nature; this definition is called naturalist definitions [1], hence, the disease is an internal circumstance disturbing the natural functioning. As per the "the biostatistical theory of disease," if a mental or bodily working condition is decreasing below threshold value then there is the disease [2].

4.1.1 Disease

In humans, disease is often used more broadly to refer to any condition that causes pain, dysfunction, distress, social problems, or death to the person afflicted, or similar problems for those in contact with the person. In this broader sense, it sometimes includes injuries, disabilities, disorders, syndromes, infections, isolated symptoms, deviant behaviors, and atypical variations of structure and function, while in other contexts and for other purposes these may be considered distinguishable categories. Diseases can affect people not only physically, but also mentally, as contracting and living with a disease can alter the affected person's perspective on life [3].

As per the definitions of normative, disease is a social convention. The amount of harm decides the level of disease, like biological or mental phenomena, therefore it is called normativist [4]. In the context of diseases, research groups argue whether caffeine-induced insomnia, baldness, freckles, and sorrow count as diseases, these are conflicts between the professionals and research interest groups. In the same way, people in society can believe something to be a disease, although this may conflict with professionals and personal interests. Another type of disease "autoimmune disease" (Ads) is a condition in which the body's immune system accidentally attacks the body itself (abnormal immune response). Generally, the function of the immune system is to protect against microbes, such as viruses and bacteria. If it is sensed by a strange intruder, it sends a signal to the combat cells to attack them. Typically, the immune system function can identify the difference between stranger cells and the cell itself [5,6].

4.1.1.1 Autoimmune diseases

ADs are chronic circumstances that characterize a heterogeneous group of disorders disturbing immunological tolerance to specific target organs and self-antigens as well as several other systems [7]. Autoimmune disease occurs when the behavior of the natural defense system of the body is not able to identify between the foreign cells and own cells, causing the body to mistakenly attack the normal cells [8] (Fig. 4.1).

There are more than 80 types of autoimmune diseases which affect a broad range of body parts. The symptoms in the early stage of ADs are similar, such as

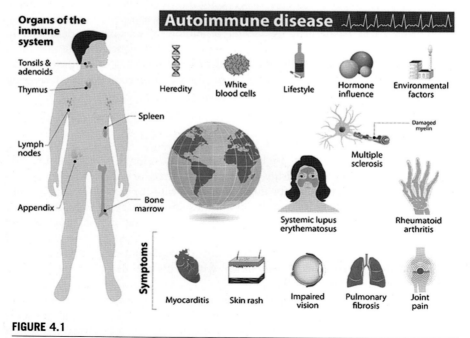

FIGURE 4.1

Symptoms of autoimmune diseases.

4.1 Introduction 69

swelling and redness, low-grade fever, trouble concentrating, numbness and tingling in the hands and feet, achy muscles, hair loss, skin rashes, and fatigue [9].

4.1.1.2 Classification of diseases

Broadly we can classify diseases under the following mental diseases, infectious diseases, physical diseases, degenerative diseases, social diseases, self-inflicted diseases, deficiency diseases, and inherited diseases. Again as per their characteristics, diseases can be categorized into acute diseases and chronic diseases. If a disease is sudden and lasts for a short time then is classed as an acute disease, and when the effects of the disease can last for months/years then it is a chronic disease [10] (Table 4.1).

Table 4.1 Classification of diseases.

Sr. no.	Category	Cause	Description	Example
1	Physical	Damage to the body	Temporary or permanent damage to any parts of the body	Leprosy, broken leg.
2	Mental	Transforms to the mind, probably with a physical cause.	Whichever disease that affects a person's mind. Sometimes a visible degeneration of brain tissue is present	Alzheimer's, dementia, schizophrenia.
3	Deficiency	Unbalanced or insufficient diet.	Nutritional disease cause by an unbalanced/inadequate diet. This may lead to starvation.	Scurvy
4	Inherited	Faulty genes	A disease comes from parents to their children by a faulty allele.	Cystic fibrosis
5	Social	Social environment or behavior	When a person's life affects their health, exposing or protecting them from certain diseases.	Hypothermia
6	Infectious	Organisms invading the body	Pathogens causing disease within the body, can be transmitted from person to person	Malaria
7	NonInfectious	Not pathogens	Any disease that is not caused by a pathogen.	Stroke
8	Degenerative	Gradual decline in function	Repair mechanisms failing, immune system begins to attack itself.	Alzheimers, strokes, cancers.
9	Self-inflicted	Self	Willful damage to own person's body by themselves.	Attempted suicide

FIGURE 4.2

The relationship among the concepts of illness, sickness, and disease.

4.1.1.3 Concept of diagnosis and treatment

Distinguishers of health depend on separating the diseased from non-diseased and the threshold value helps patients to recover [4]. If the threshold value is too small, the patient can be diagnosed, underdiagnosed, and be caused unnecessary uncertainty, anxiety, pain, and death. Another ethical issue which is very important for the diagnosis of diseases of the patient, in modern medicine, a wide range of tests is done to identify the diseases, which results in early treatment, sometimes due to that disease; the result of uncertainty is parsley (Fig. 4.2).

4.1.2 Medicine

A medicine is a chemical compound that halts or cures diseases as per the symptoms of sickness. Nowadays, medicines come from a wide variety of sources. Many have been developed from substances found in nature, and even today many are extracted from plants [9].

Medicine, also known as pharmaceutical drug, is a drug used to diagnose, treat, treat, or prevent diseases. Drug therapy or pharmacotherapy is a very important part of the medical field that relies on the scientific principle of pharmacy for proper management [11–14].

The medicine or drug is a substance which is swallowed in the form of tablet capsules, injected inside the body, or applied outside the body, due to which the body gradually starts to perform normal functioning [15]. Mainly drugs are used for the treatment of diseases or to cure abnormal circumstance, such as antibiotics used to cure an infection. Some medicines treat depression and can be given to relieve symptoms of an illness like painkillers for reducing the pain. Vaccinations are a treatment to prevent diseases such as flu [15].

4.1.2.1 Medicine working

Various techniques are found in the human body and this path is called the "route." This route can be in the form of capsules, tablets, or liquids or by other means [16,17] (Fig. 4.3).

4.1 Introduction 71

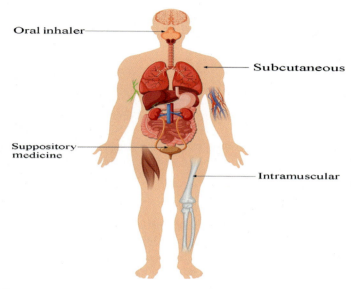

FIGURE 4.3

Routes for medication.

Routes for medicine can be oral, nasal (into the nose), buccal (placed in the cheek), sublingual (placed under the tongue), eye- and eardrops, and transdermal (through the skin) in the human body [18].

4.1.2.2 Different types of medicines

Some drugs are used to treat cancer by killing cells or they prevent cell division or multiplying. Some medicines correct low levels of natural body chemicals such as certain hormones or vitamins, some can also affect parts of the nervous system that control the body's process. Antibiotics fight bacterial infections by working to kill bacteria or prevent their multiplication so that the body's immune system can fight infection well [19,20].

When the body does not make sufficient levels of hormones, a person can become ill. For example, diabetes has a dependence on insulin (the hormone that regulates the amount of glucose in the body) and the pancreas is unable to produce a sufficient amount of insulin, whereas other patients have thyroid hormone problems. In such a case medicine is given for the correct production of hormones [21].

4.1.2.3 Discovering new medicines

The drug is determined using preclinical testing to determine how to develop it for its intended use. The purpose is to observe how the drugs are distributed and

FIGURE 4.4

Seqential steps for discovering a new medicine.

absorbed in the body, and how they are broken down and removed from the body [22,23] (Fig. 4.4).

Clinical development involves testing these drugs on human volunteers to provide more information about those identified as clinical trials, as well as their safety and effectiveness [24,25].

4.1.2.4 Role of intelligent algorithm

Typically, AI algorithms carry out tasks that require human intelligence to complete, such as pattern and speech recognition, image analysis, and decision-making. In short, AI algorithms can improve humans to automate the toughest tasks they are trained by algorithms to do.

The science of making intelligent computer programs is artificial intelligence (AI). The aim of AI is to help doctors and patients in clinical diagnosis and treatment, and to enhance the quality of therapy and reduce the rate of error. The most frequently used AI methods are fuzzy logic, genetic algorithms, artificial neural networks (ANNs), and expert systems (ES). The main function of ES estimates the cause and effect relationship with patient data and gives recommendations to the doctor. The neurons in the ANN mimic the biological nervous system [26].

4.1.2.5 History of AI

The meaning of intelligence is to solve a problem and the knowledge of logic, self-awareness, creativity, and understanding. Basically when we talk about intelligence then we assume human or animal intelligence, but in both cases the joint intelligence is known as natural intelligence. Similarly, plants have also intelligence, they are not able to show it but do react to their environment [27].

Artificial intelligence is a subpart of computer science, which acts like a machine with an association of hardware and software. Here machine refers to an artificial, and the function of an artificial machine based on the observation process is cognitive.

A play named Rossumovi Univerzální Roboti (Rossum's Universal Robots: R. U.R.) was published by the Czech writer Karel Čapek in 1920, this type of game started for the first time in the world, in collaboration with the industry, created an artificial person called a robot. By the association of the R.U.R industry create an artificial people called a robot.

In 1956 the first AI workshop was held at The Dartmouth conference. At that conference, various researchers from Massachusetts Institute of Technology

4.1 Introduction

(MIT), IBM, and Carnegie Mellon University (CMU) were participated and discuss an optimistic way of developing AI.

4.1.3 History of AI in medicine

"Healthcare is an information industry that continues to think that it is a biological industry" [28]. When a machine gets trained through existing system and performs in a better way it is intelligent. In reality it is observed by plants, human, nonhumans, and animals. Intelligence lies in artificial machines [29].

In today's scenario, the active research in the field of computer science is artificial intelligence (AI) and the objective of this technology is to build a system which imitates human intelligence which can be used in various human behaviors together with medicine [30].

Artificial intelligence in healthcare refers to the use of complex algorithms designed to perform certain tasks in an automated fashion. When researchers, doctors, and scientists inject data into computers, the newly built algorithms can review, interpret, and even suggest solutions to complex medical problems [31].

In the 1960s AI's initial foray into medicine primarily began. Among the early AI works, was Stanford's Ted Shortliffe and the innovative program he made, named "MYCIN," which worked on a basic rule-based expert system through if—then regulation with conviction values to suggest different results for infectious diseases [32,33].

In 1982 Szolovits edited a textbook on the title "artificial intelligence in medicine" to proved that MYCIN is superior to human infectious disease experts [33]. At the beginning of this time, various institutions such as MIT, Stanford, Pittsburgh, and Rutgers, as well as some institutions in Europe, started participating in "AI in Medicine." In Marseille, France in 1987, the bi-yearly meeting started on Artificial Intelligence in Medicine (AIME) [30]. Today AI has been used in a broad of range of medical fields like medicine, healthcare, diagnosis, etc. The objectives of the use of AI in medicine are [34]:

1. Assessment of risk in assessing the onset of disease and the success of treatment.
2. Supervising or reducing complications at the right time.
3. An important role in the care of current patients.
4. In the study of permanent pathology with treatment utility research.

In the digital future, the patient can see the symptoms of their disease in the computer before doctors through AI practices in medicine, with the help of AI technology; treatment can go for root causes rather than direct diagnosis [35].

AI technology has been widely used in the medical sector [36–39], manufacturing sector, information-communication industry [40], and information-communication industry. It is beginning to be the most used and commercialized technology in the medical field [41,42]. As per the definition of AI, it performs typical tasks for which it needs human-level intelligence. Machine learning is a

74 **CHAPTER 4** Artificial itelligence in medicine

subpart of AI which performs in an efficient manner and gives fast and accurate results through the use of statistical algorithms and complex computing [43,44].

Various complex algorithm techniques used by AI follow human cognition for the investigation of huge complex data sets and the beauty of AI is that it can give a conclusion without the interaction of humans [45].

Artificial Intelligence is just like a machine, which performs human tasks with a guaranteed degree of intelligence. Such a type of process may include self-rectification, learning, and interpretation [27,46]. In 1950, John McCarthy introduced the phrase "artificial intelligence" and he said that the machine can simulate all the tasks in future, such as problem-solving and create abstraction, which are held in reserve for humans today. As per Stuart Russell and Peter Norvig, AI is "the study of agents that receive percepts from the environment and perform actions" [47]. There are two concepts, the first being covenants through thought process and logic, and the second through behavior. In thought processes, two important areas are human thinking, thinking rationally [48].

Artificial intelligence can be classified into two categories: weak artificial intelligence and strong artificial intelligence [49,50]. Weak artificial intelligence systems are designed in such a way that they only do a specific task, whereas strong artificial intelligence system respond like a human, such as a self-driving car.

The definition of AI has changed in the modern era; it can be defined in techno-modern terms as an intelligent environment which is related to the new paradigm design of artificial intelligence. An artificial intelligence agent has to incorporate some basic concepts for the design of new AI, which are described below:

1. Reasoning
2. Learning
3. Problem solving
4. Language understanding
5. Perception

1. Learning

   Learning is the fundamental foundation for the improvement of any new technique, which is based on a heuristics approach and it validates the facility for the new agents to always enhance the learning system which overcomes the existing system (Fig. 4.5).

2. Reasoning

   In real time, a quick response and correct decision is always an expectation from agents and is dependent on the logic and situation of the particular task. One of the challenges is that AI is not fully capable of distinguishing between relevant and irrelevant.

3. Perception

   In the case of observation, perception is improved by analyzing physical and virtual relations of objects, as it is based on continuous learning and learning and improvement relations of object characteristics.

FIGURE 4.5

Actions of artificial intelligence.

4. Problem solving

The main collective issue of AI is problem solving because the classification of problems is a tremendous assignment, so infrequent problems can be classified as either general or specific. A tailor-made technique approach is made under a special-purpose mechanism, while in general techniques problem-solving is sequences of stages in a cascaded manner.

5. Language perceptive

Basically a language is a measurement of information consumed through convention, but the language may be in various appearances which distinguish another language from the human speaking mechanism [51].

4.1.4 Drug discovery process

Discovery of a new medicine is a critical and expensive process; various complex diagnostic processes are needed for developing any new medicine. By using machine learning it can be made very easy and efficient; through this technique it is possible to cut the time and investment. For drug development, there are four important stages where AI is already used [52].

1. Stage 1: identify goals for involvement
2. Stage 2: discovering drug users
3. Stage 3: speeding up medical experiments
4. Stage 4: discovery biomarkers for diagnosing the disease

Stage 1: Identify goals for involvement:

The understanding of biological sources and their resistance techniques are the initial stage of drug identification for the diseases. Most of the time

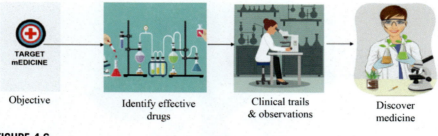

FIGURE 4.6

Drug discovery process.

proteins are the basic needs for the treatment of diseases. With the availability of high-throughput mechanisms, for instance, deep sequencing and shRNA (short hairpin RNA) can collect and discover feasible paths. But still there is a challenge in traditional techniques to integrate different source of parameters. Analysis for the available data can use machine-learning algorithms in more efficient ways to automatically identify a target protein (Fig. 4.6).

Stage 2: Identify prospective drug:

In the second stage of the drug discovery, focus on identifying the compound with an association of target molecule in a preferred way through various types of screening for their outcome on potential drug compounds, as well as realizing the impact of toxicity, such as the type of compound may be synthetic or bioengineered or natural. However, the existing system may be inaccurate due to false-positive results. So various algorithms are used to get the narrow result, and also take a long time for computation to identify the best prospective drug. Machine learning helps in the prediction of the right drug with the correctness of the compound's molecule based on molecular descriptors and structural fingerprints. The outcome is based on minimal toxicity or side effects. So AI can save huge time in the design of the drug.

Stage 3: Momentum of clinical trials:

In general, it is very difficult to choose a correct candidate for trials, because if the choice is a mistake then it extend the cost of the trial, which includes the costing of the time and resources, Machine learning operates in such a way to automatically recognize appropriate aspirants for clinical trials, thus ensuring the real-time correct result. The algorithm of machine learning facilitates recognition of the right pattern. It also assists in the detection of the wrong pattern, when the result is irrelevant or not up to the mark.

Stage 4: Discover biomarkers for diagnosing the disease:

The patient is the only entity that is used to diagnose research treatments, thus some techniques are very expensive because their participation requires inexpensive laboratory equipment as well as expert knowledge of genome sequencing. A biomarker is basically a molecule found in human blood that assures that the disease is not present in humans after diagnosis; such types of diagnosis are safe and inexpensive [51].

For a disease, identification of appropriate biomarkers is very hard, but AI can computerize a large section of the existing manual diagnosis.

4.1.5 Machine-learning algorithms in medicine

There are various machine learning algorithms to extract the disease characteristics of the patients, who record all these data by knowing the symptoms of existing diseases, past history of any chronic diseases, as well as environmental lifestyle, etc. Finally, an appropriate medication can be provided to the patient. For this purpose, some relevant algorithms are described here.

1. **Regression Algorithms**
   a. Linear regression
   b. Logistic Regression
2. **Instance-based Algorithms**
   a. Support Vector Machine
   b. k-Nearest Neighbor (kNN)
3. **Decision Tree Algorithms**
   a. Decision Tree
4. **Bayesian Algorithms**
   a. Naïve Bayes
5. **Clustering Algorithms**
   a. K-Means Algorithm
6. **Artificial Neural Network Algorithms**
   a. Perceptron
   b. ANN
7. **Deep Learning Algorithms**
   a. Convolutional Neural Network
   b. Recurrent Neural Network
8. **Ensemble Algorithms**
   a. Random Forest
9. **Other Machine Learning Algorithms**
   a. Natural Language Processing
   b. Neural Network
   c. Word Vectors
   d. Term frequency-inverse document frequency. TF-IDF
   e. Hidden Markov Model (HMM)

Here some of the important machine algorithms which help to understand the concept of medicine are described below.

78 CHAPTER 4 Artificial itelligence in medicine

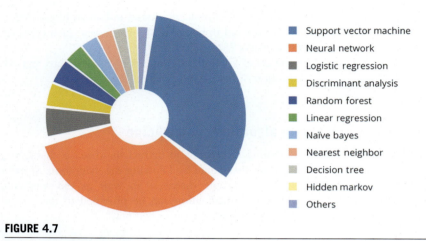

FIGURE 4.7

Result of AI algorithms by sciforce.

4.1.5.1 Linear regression

The regression algorithm is a subset of machine learning algorithms and a super subset of AI. The main function of this algorithm is to predict the new target output, which is based on the model of evaluation of the relationship between the input and the target output. The model is based on the training and testing data. In a linear regression algorithm, regression refers to modeling to predict the target value (Y) through the target function (f), which is based upon independent variables with a linear relation between dependent (Y) and independent variables (X), i.e. $Y = f(X)$, so this is called predictive analysis to achieve the most accurate possible predictions (Fig. 4.7).

4.1.5.2 Logistic regression

Logical regression is predictive analysis and gives a binary result (Y) 0 or 1, and it was developed in 1958 by the statistician David Cox. In the case of a correct medicine, the main objective of this algorithm is to find out the logical-mathematical expression for real-time medicine prediction (Y) when all the independent variables are known. It predicts the target output if the given "mass of tissue" is malignant or benign (Fig. 4.8).

4.1.5.3 Support vector machine

SVM is a classifier and separates clusters through the hyperplane. As per the given labeled training data of supervised learning, the outcome of this algorithm is an optimal hyperplane, where each class is situated side by side of this plane in the case of two dimensions (Fig. 4.9).

This algorithms uses both challenges, either regression or classification, with respect to the features (denoted by n) of the data. It can also plot in n-dimensions

4.1 Introduction 79

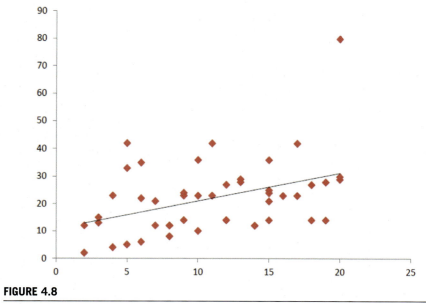
FIGURE 4.8
Linear regression.

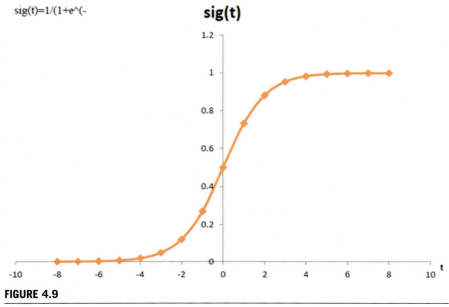
FIGURE 4.9
Mathematical function of logistic regression.

space, the value of each feature is positioned in space with reference to the coordinate (Fig. 4.10).

4.1.5.4 Convolutional neural network

The important algorithm in AI for medicine is the convolutional neural network, which has lots of features that are very fruitful in the "medicine" field. Various raw data are very useful for unknown feature extraction, each hidden layer having a random weight with the engagement of each neuron. Such a configuration slows down the training model, and after learning these relationships can give an approximate correct output (Fig. 4.11).

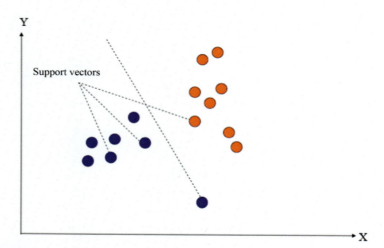

FIGURE 4.10

Visualization of support vector machine.

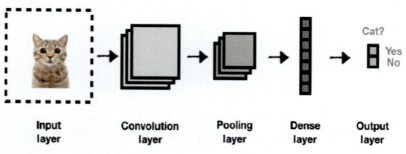

FIGURE 4.11

Convolutional neural network (CNN).

4.1 Introduction

Similar to neural networks, the convolutional neural network (CNN) consists of weighted neurons, where each neuron is weighted randomly and, after calculation, the algebraic values of the neurons are transferred to the next layer through an activation function. In the output function, the whole network acts like a "loss function", CNN has the following computational steps, which described below:

1. Convolution Layer
2. ReLU Layer
3. Pooling Layer
4. Fully Connected Layer
5. AIM in Current Scenario

Convolution Layer:

A very unique features of Convolution is the translational invariant; natural filtration of convolution is a characteristic feature of interest, where CNN algorithms study the features, which consist of the **resulting reference**. The following steps are involved in convolution [53]:

- **Line up** the feature and the image.
- **Multiply** each **image** pixel by corresponding **feature** pixel.
- **Add** the values and find the **sum**.
- **Divide** the sum by the **total** number of pixels in the **feature**.

Rectified Linear Unit (ReLU Layer):

It is a activation function, and if the values of input cross the threshold value then it will be active and transform the value. When input value is below zero then output is zero, and also if the value of input rises at the highest value then give the output as per the linear mathematical relationship (Fig. 4.12).

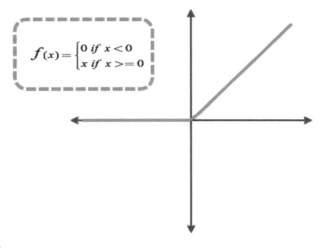

FIGURE 4.12

Mathematical function of ReLU layer.

82 CHAPTER 4 Artificial itelligence in medicine

Pooling Layer:

In this case, when the image stack has been shrinking into a smaller size, in this sequence when the input has passed in the activation layer then pooling is done, therefore four steps are required.

- Pick a window size (usually 2 or 3).
- Pick a stride (usually 2).
- Walk your window across your filtered images.
- From each window, take the maximum value.

After compilation of these steps comes the output, which will become an input for the next layer "Fully Connected Layer."

Fully Connected Layers:

In the CNN algorithm, the "Fully Connected Layer" is the last layer in this sequence; the association of this layer is the neurons of the preceding layer. Through each neuron in the next layer being connected, this layer consists of logic at a high level with the linkage of all its paths from initial level input to the final level, which leads to true and accurate results.

AIM in Current Scenario:

In the current scenario, machines are helping human intelligence and patients, in 1988, Hong told the possibility of AI practice in medicine with new technology for testing, through the necessary training, decision making, and research and development possible.

The main objective of artificial intelligence is to provide doctors with the best diagnosis for better treatment along with reducing the rate of medical error. The techniques used by AI for drugs are genetic algorithms, expert systems (ES), artificial neural networks (ANNs), and fuzzy logic. Expert systems take a lot of input data from patients and make the right recommendations to the doctor in real time. The role of fuzzy systems is to take uncertain drug data and generate scientific manifestations and predictable results. The biological nervous system is basically neurons that are associated with ANN and all these techniques give direct benefit to patients.

4.1.6 Expert systems

An expert system is the most significant application, which imitates the quality of expertise in a specialization area. An expert system has the ability to perform the same as a specialist, in planning, designing, controlling, recommendations, generalizing, interpreting, and diagnosing. The most important features of expert systems are inference mechanism and the database that make it superior to other existing support systems. The traditional program consists of the database along with algorithms, where the expert system includes of database with an inference mechanism. In the case of an expert system, the database is an essential part.; it contains data, rules, relationships, information, problem definition, solutions, and information for how to approach a problem. The engine is a core part of an expert system [54]. It provides a policy of sophisticated data and databases and communicates to the user and enables solutions that can be achieved through the logical interpretation of the system.

The role of an expert system for medicine is to provide an accurate result with respect to the appropriate structural questions. The objective of AI is never to replace the existing physician but to help/advise them in their work and lead to the correct decisions based on the patient's data.

A well-known expert system is "MYCIN" built-up at Stanford University, for the treatment and diagnosis of bacterial infectious diseases and decrees the uses of antibiotic. It takes the outcome of laboratory results, patient data, diagnosis report, and functions of treatment planning. It is a basically tool which details to a patient's doctor the diseases which are identified in the blood by bacteria. For a response/decision the doctor must answer all the questions asked by the MYCIN, but despite this it has never been used in real practice due to ethical issues.

4.1.7 Fuzzy expert systems

In classical set theory based on fuzzy logic and its subsets tell us which data is imprecise or vague, and decision through logical variations within the framework through fuzzy logic rules.

The degree of membership of entities can range from 0 to 1, as well as the gray level used to avoid the true—false dilemma, to overcome the challenges which arise from working with uncertain data and symbolic logic results. Most of the concepts in the field of medicine today are unclear, which is not suitable for medical applications; the uncertain medical need is met by a fuzzy set that contradicts the logic of how to generate solutions with predictive results.

4.1.8 Artificial neural networks

This is a nonlinear model of a complex structure, where independent variables are an input for the system which are coupled with a dependent variable to provide the output of the system. The power in artificial neural networks (ANN) is drawn from its parallel structure and generalized learning due to its skill and it can solve complex problems. Generalization is one way that ANN constructs appropriate responses to inputs not collected in the education or learning process. The basic unit of the main process of ANN is the neurons which are nonlinear. Consequently, the outcome of combination of the cells spread all over the network is also nonlinear. Due to its complex nonlinear structure, it can solve very complex problems in an easy way in medical practice (Fig. 4.13).

84 CHAPTER 4 Artificial itelligence in medicine

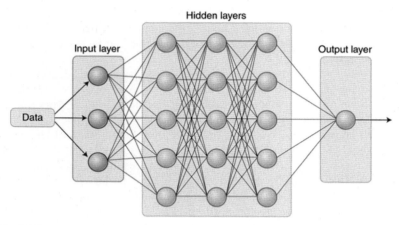

FIGURE 4.13

Layer in artificial neural networks.

4.2 Conclusion

In the current scenario, technologies of AI are being trialed in the medical field for discovery of new medicines, disease identification, chronic conditions of patients, and services in healthcare. These technologies have the ability to facilitate important medicinal issues, but gain strength be restricting the quality of digital health data and the absence of the several human characteristics. AI also raises social and ethical issues, and in many cases these issues overlap with the other raised issues such as healthcare technologies and medical data, In the field of medicine, the role of AI is subject to the regulation. There is a need to develop AI in such a way that it should be compatible and transparent with the public interest. AI is not able to mark the separation between ethical and social issues with uses of automation and data uses. Trust in the technologies have various parameters and there are challenges which arise by the use of assistive technologies. AI controls medical equipment for treatment and gives its verdict in the medical field, where the main key issues are safety and reliability, because if any error comes and AI does not have the solution then it may have serious implications. AI prediction relies on raw data, so if future stability quality declines, significant accuracy of results will also be a practical challenge. In the current scenario medical records are not in digitized form and also have a lack of standardization, data labeling, and interoperability. AI cannot easily replicate doctors, because very complex verdicts are involved in clinical practice, such as human knowledge combined with contextual knowledge that cannot be achieved through machines.

References

[1] P.S. Davies, Norms of nature: naturalism and the nature of functions, MIT Press, 2003.

[2] C. Boorse, On the distinction between disease and illness, Philos. Public. Aff. (1975) 49–68.

[3] L. Castano, G.S. Eisenbarth, Type-I diabetes: a chronic autoimmune disease of human, mouse, and rat, Annu. Rev. Immunol 8 (1) (1990) 647–679.

[4] D. Album, S. Westin, Do diseases have a prestige hierarchy? A survey among physicians and medical students, Soc. Sci. Med. 66 (1) (2008) 182–188.

[5] J. Dalmau, C. Geis, F. Graus, Autoantibodies to synaptic receptors and neuronal cell surface proteins in autoimmune diseases of the central nervous system, Physiol. Rev. 97 (2) (2017) 839–887.

[6] Stacy Sampson, 2019, Autoimmune-disorders. <https://www.healthline.com/health/autoimmune-disorders>.

[7] J.M. Anaya, Common mechanisms of autoimmune diseases (the autoimmune tautology), Autoimmun. Rev. 11 (11) (2012) 781–784.

[8] M. Jangra, S.K. Dhull, K.K. Singh, ECG arrhythmia classification using modified visual geometry group network (mVGGNet), J. Intell. Fuzzy Syst. (2020) 1–15.

[9] Ana-Maria Orbai, Symptoms of Autoimmune Disease, 2019, <https://www.hopkinsmedicine.org/health/wellness-and-prevention/what-are-common-symptoms-of-autoimmune-disease>.

[10] Wikibooks, 2019, A-level biology/human health and disease/introduction. <https://en.wikibooks.org/wiki/A-level_Biology/Human_Health_and_Disease/introduction>.

[11] C.P. Adams, V.V. Brantner, Estimating the cost of new drug development: is it really $802 million? Health Aff. 25 (2) (2006) 420–428.

[12] P.P. George, J.A. Molina, J. Cheah, S.C. Chan, B.P. Lim, The evolving role of the community pharmacist in chronic disease management-a literature review, Ann. Acad. Med. Singap. 39 (11) (2010) 861–867.

[13] D.N. Sarma, M.L. Barrett, M.L. Chavez, P. Gardiner, R. Ko, G.B. Mahady, et al., Safety of green tea extracts, Drug. Saf. 31 (6) (2008) 469–484.

[14] A. Vlietinck, L. Pieters, S. Apers, Legal requirements for the quality of herbal substances and herbal preparations for the manufacturing of herbal medicinal products in the European Union, Planta Medica 75 (07) (2009) 683–688.

[15] BDS, Medication administration curriculum section II 2011, <https://www.dhhs.nh.gov/dcbcs/bds/nurses/documents/sectionII.pdf>.

[16] M.K. Giacomini, D.J. Cook, Users' guides to the medical literature: XXIII. Qualitative research in health care B. What are the results and how do they help me care for my patients? Evidence-based medicine working group, Jama 284 (4) (2000) 478–482.

[17] A. Mehrotra, S. Tripathi, K.K. Singh, P. Khandelwal, 2014, Blood vessel extraction for retinal images using morphological operator and KCN clustering. In 2014 IEEE international advance computing conference (IACC) (pp. 1142–1146). IEEE.

[18] S.I. Tomarev, Eyeing a new route along an old pathway, Nat. Med. 7 (3) (2001) 294–295.

[19] M. Jangra, S.K. Dhull, K.K. Singh, Recent trends in arrhythmia beat detection: a review, in: B.M. Prasad, K.K. Singh, N. Ruhil, K. Singh, R. O'Kennedy (Eds.), Communication and computing systems, 1st ed., 2017, pp. 177–183.

[20] J. Kennedy, I. Tuleu, K. Mackay, Unfilled prescriptions of medicare beneficiaries: prevalence, reasons, and types of medicines prescribed, J. Manag. Care Pharm. 14 (6) (2008) 553−560.

[21] Elora Hilmas, 2018, Understanding medicines and what they do, <https://kidshealth.org/en/teens/meds.html>.

[22] S.S. Mondal, N. Mandal, A. Singh, K.K. Singh, Blood vessel detection from retinal fundas images using GIFKCN classifier, Procedia Comp. Sci. 167 (2020) 2060−2069.

[23] AnaBios, 2019, Enabling human-focused drug discovery, <https://anabios.com/ana-bios-advantage/>.

[24] J.H. Lee, E. Kim, T.E. Sung, K. Shin, Factors affecting pricing in patent licensing contracts in the biopharmaceutical industry, Sustainability 10 (9) (2018) 3143.

[25] Yourgenome, 2017, Yourgenome, <https://www.yourgenome.org/facts/how-are-drugs-designed-and-developed>.

[26] H.B. Hlima, T. Bohli, M. Kraiem, A. Ouederni, L. Mellouli, P. Michaud, et al., Combined effect of spirulina platensis and punica granatum peel extacts: phytochemical content and antiphytophatogenic activity, Appl. Sci. 9 (24) (2019) 5475.

[27] Lasse Schultebraucks, 2018, A short history of artificial intelligence. <https://dev.to/lschultebraucks/a-short-history-of-artificial-intelligence-7hm>.

[28] E.H. Shortliffe, The adolescence of AI in medicine: will the field come of age in the '90s? Artif. Intell. Med. 5 (2) (1993) 93−106.

[29] D. Saranya, M. Phil, A study on artificial intelligence and its applications, Int. J. Adv. Res. Comp. Commun. Eng. 5 (4) (2016) 313−315.

[30] L.Q. Shu, Y.K. Sun, L.H. Tan, Q. Shu, A.C. Chang, Application of artificial intelligence in pediatrics: past, present and future, World J. Pediatr. 15 (2019) 105−108.

[31] S. Yung, EqOpTech Publications, 2018.

[32] E.H. Shortliffe, R. Davis, S.G. Axline, B.G. Buchanan, C.C. Green, S.N. Cohen, Computer-based consultations in clinical therapeutics: explanation and rule acquisition capabilities of the MYCIN system, Comp. Biomed. Res. 8 (4) (1975) 303−320.

[33] P. Szolovits, W.J. Long, The development of clinical expertise in the computer, Artif. Intell. Med. (1982) 79−117.

[34] A.N. Ramesh, C. Kambhampati, J.R. Monson, P.J. Drew, Artificial intelligence in medicine, Ann. R. Coll. Surg. Engl. 86 (5) (2004) 334.

[35] G. Daniel, 2019, AI in medicine. <http://sitn.hms.harvard.edu/flash/2019/artificial-intelligence-in-medicine-applications-implications-and-limitations/>.

[36] D.D. Miller, E.W. Brown, Artificial intelligence in medical practice: the question to the answer? Am. J. Med. 131 (2) (2018) 129−133.

[37] S. Oh, J.H. Kim, S.W. Choi, H.J. Lee, J. Hong, S.H. Kwon, Physician confidence in artificial intelligence: an online mobile survey, J. Med. Internet Res. 21 (3) (2019) e12422.

[38] L. Ohno-Machado, Research on machine learning issues in biomedical informatics modeling, J. Biomed. Inf. 37 (4) (2004) 221−223.

[39] V.L. Patel, E.H. Shortliffe, M. Stefanelli, P. Szolovits, M.R. Berthold, R. Bellazzi, et al., The coming of age of artificial intelligence in medicine, Artif. Intell. Med. 46 (1) (2009) 5−17.

[40] S. Pyo, J. Lee, M. Cha, H. Jang, Predictability of machine learning techniques to forecast the trends of market index prices: hypothesis testing for the Korean stock markets, PLoS One 12 (11) (2017) e0188107.

References   **87**

[41] K.J. Dreyer, J.R. Geis, When machines think: radiology's next frontier, Radiology 285 (3) (2017) 713−718.

[42] S. Nemati, A. Holder, F. Razmi, M.D. Stanley, G.D. Clifford, T.G. Buchman, An interpretable machine learning model for accurate prediction of sepsis in the ICU, Crit. Care Med. 46 (4) (2018) 547−553.

[43] M. Alsharqi, W. Woodward, A. Mumith, D. Markham, R. Upton, P. Leeson, Artificial intelligence and echocardiography, Echo Res. Pract. (2018) 30400053.

[44] K. Shameer, K.W. Johnson, B.S. Glicksberg, J.T. Dudley, P.P. Sengupta, Machine learning in cardiovascular medicine: are we there yet? Heart 104 (14) (2018) 1156−1164.

[45] F. Jiang, Y. Jiang, H. Zhi, Y. Dong, H. Li, S. Ma, et al., Artificial intelligence in healthcare: past, present and future, Stroke Vasc. Neurol. 2 (4) (2017) 230−243.

[46] K. Chethan, 2018, Artificial intelligence: definition, types, examples, technologies. <https://medium.com/@chethankumargn/artificial-intelligence-definition-types-examples-technologies-962ea75c7b9b>.

[47] S.J. Russell, P. Norvig, Artificial intelligence: a modern approach, Pearson Education Limited, Malaysia, 2016.

[48] J. Padikkapparambil, C. Ncube, K.K. Singh, A. Singh, Internet of things technologies for elderly health-care applications, Emergence of pharmaceutical industry growth with industrial IoT approach, Academic Press, 2020, pp. 217−243.

[49] J. Frankenfield, 2020, Artificial intelligence (AI), <https://www.investopedia.com/terms/a/artificial-intelligence-ai.asp>.

[50] M. Singh, S. Sachan, A. Singh, K.K. Singh, Internet of things in pharma industry: possibilities and challenges, Émerg. Pharm. Ind. Growth Ind. IoT Approach (2020) 195−216.

[51] S. Tripathi, K.K. Singh, B.K. Singh, A. Mehrotra, Automatic detection of exudates in retinal fundus images using differential morphological profile, Int. J. Eng. Technol. 5 (3) (2013) 2024−2029.

[52] Markus, 2019, AI for diagnostics, drug development, treatment personalisation, and gene editing. <https://www.datarevenue.com/en-blog/artificial-intelligence-in-medicine>.

[53] S.K. Dhull, K.K. Singh, ECG beat classifiers: a journey from ANN To DNN, Procedia Comp. Sci. 167 (2020) 747−759.

[54] Prasad, B.M.K., C.S. Singh, and K.K. Singh. Brain wave interfaced electric wheel-chair for disabled & paralysed persons (2016): 773−776.

CHAPTER 5

Diagnosing of disease using machine learning

Pushpa Singh[1], Narendra Singh[2], Krishna Kant Singh[3] and Akansha Singh[4]

[1]*Department of Computer Science and Engineering, Delhi Technical Campus, Greater Noida, India*

[2]*Department of Management Studies, G. L. Bajaj Institute of Management and Research, Greater Noida, India*

[3]*Faculty of Engineering & Technology, Jain (Deemed-to-be University), Bengaluru, India*

[4]*Department of CSE, ASET, Amity University Uttar Pradesh, Noida, India*

5.1 Introduction

In current years, substantial efforts have been made for the development of computer-based disease diagnosis applications. Machine learning (ML) has proven very effective in the diagnosis and prediction of fatal diseases. Machine learning has transformed the conventional health sector into a smart and intelligent health sector. Machine learning is an artificial intelligence (AI)-based technique. AI-based techniques assist in investigating medical data more accurately and quickly and enable doctors to be more precise with correct and timely diagnoses. Diagnosis is a process for recognizing and understanding the nature of a disease. AI has become more accurate for the diagnosis and prognosis of fatal diseases through available clinical data. These data are available in various forms like demographics, medical notes, electronic health records (EHR), physical checkups, and medical imaging [1]. Medical imaging consists of X-rays, CT scans, MRIs, ultrasound, nuclear medicine imaging, mammography, etc. [2]. ML becomes a more practical tool to provide diagnostic information for the various diseases.

At present the healthcare industries rely on computer technology. Computers are used to collect an enormous amount of data, also known as the big data of patient disease, treatments, and outcomes. One can automatically achieve valuable data improvements for the efficiency of treatments, or relations between side effects, match new patterns, and diagnose thee disease of a new patient based on old data using ML techniques. ML techniques are helpful to recognize and detect health anomalies at the earliest stage. Precise diagnosis is very significant for determining the correct therapy, medicine, or treatment. ML methods can be easily applied to the clinical records to create a descriptive study of clinical features. A ML algorithm is extensively used in the diagnosis of several diseases like

Machine Learning and the Internet of Medical Things in Healthcare. DOI: https://doi.org/10.1016/B978-0-12-821229-5.00003-3
Copyright © 2021 Elsevier Inc. All rights reserved.

90 CHAPTER 5 Diagnosing of disease using machine learning

neurological, cardiovascular, cancer, diabetic, etc. ML is capable of diagnosing and detecting a disease at an earlier stage that assists in reducing the number of readmissions to hospitals, the cost of expensive treatment, and also increases the survival rate in cases of fatal disease.

AI and ML applications in the field of health care aren't just limited to diagnosing a disease, but also offer the possible combination of the treatment. The aim of machine learning systems is not to substitute for the doctors, but to support the decisions and make recommendations based on their analysis of the patients [3]. Major diseases where machine learning tools are remarkable include cancer, neurological disease, and cardiovascular disease. The initial diagnoses are vital to avoid the decline of patients' health conditions. Moreover, early diagnoses could be possibly attained through refining the investigative measures on images, genetics, Electronic Medical Records, etc., which is the power of the AI techniques [1].

Cancer, neurological, and cardiovascular-disease are life-threatening diseases that need to be diagnosed and treated in a timely manner. ML is used for the classification of objects, such as a lesion into certain classes like abnormal or normal, and lesions or nonlesions [4]. In diseases like cancer, where there is no specific reason for their occurrence, early detection and proper diagnosis systems can save lives. A mammogram may find a cancer as long as 2 or 3 years before it can be felt. Convolutional neural network (CNN), a class of deep learning system, can ultimately read mammograms that outperform compared to the traditional CAD system [5]. CNN has the potential to transform medical image analysis applied in a particular field of mammography breast cancer diagnosis [6]. A multistage classifier comprises support vector machine (SVM), Decision Tree (DT), Naive Bayes (NB), Linear Discriminant Analysis (LDA), Logistic Regression (LR), and K-nearest neighbor (KNN), are used for the diagnosis, classification, and prediction of various diseases.

This chapter elaborates the role of machine-learning techniques in the healthcare industry to diagnose and predict major diseases. SVM, KNN, NB, and DT are discussed with regard to the diagnosis of disease. A basic conceptual diagnostic model for the classification of several diseases is discussed in this chapter. The model is trained and tested based on Python to understand the case study of breast cancer diagnosis. A confusion matrix is used to study the performance metrics of the classifier.

5.2 Background and related work

The traditional diagnosis system is poorly designed systems, and a lack of integration between various equipment and communication systems across the health sectors and within an individual healthcare system. Such conventional systems could damage patients' health and prevent the provision of relevant treatment due to a lack of advance diagnostic systems. There were the following challenges in the conventional healthcare system.

5.2.1 Challenges in conventional healthcare system

1. The traditional diagnosis system is very long and the patient was asked about a wider range of questions than just their symptoms.
2. The traditional diagnosis system depends solely on individual knowledge and skill of the practitioner.
3. Lack of quick data processing and decision-making capability could mislead the diagnosis system.
4. Inadequacy in evaluation of large amounts of complex healthcare data available in multimedia form.
5. The low quality of the traditional digital image processors produces complexity and inefficiency in order to analyze the medical image.
6. The traditional diagnosis system was a time-consuming process.
7. Lack of modern tools, techniques, high-speed network, processor, storage, and computing power.

Recent advancementa in telecommunication systems [7], computation power [8], big data [9], and Internet of things (IoT) [10] have been widely used in the healthcare sector [11]. Recent healthcare applications have been empowered by AI and consequently have become smart and intelligent health care by offering accurate disease diagnostic systems and prediction techniques and removing the barrier of the conventional healthcare system. Integration of ML and other related techniques must require several points of diagnosis and decision-making in the healthcare system [12]. ML provides a dominant method for developing sophisticated, automatic, and objective algorithms for the diagnosis of disease [13−15]. The conventional health care system had greater diagnostic errors and unsystematic treatment plans. Diagnostic errors have been linked to as many as 10% of all patient deaths and may also account for between 6% and 17% of all hospital complications [16]. ML algorithms are studied for the diagnosis of diverse diseases, for example, heart problems [17], diabetes disease [18], liver disease [19,20], dengue disease [21], and hepatitis disease [22] etc. Neurological and cardiovascular diseases and cancer are some life-threatening diseases which always require the most appropriate diagnostic system at an early stage. This chapter outlines recent algorithms and tools of ML which have made substantial impacts in the diagnosis and prediction of several diseases.

5.2.2 Machine-learning tools for diagnosis and prediction

ML is a branch of AI, which basically emphasizes making computers intelligent and able to take decisions in critical situations. ML programming tools are key for the success of a machine-learning project in health care. ML is one of the potential solutions to diagnostic challenges, mainly when applied to image recognition in oncology (e.g., cancer tests) and pathology (e.g., bodily fluid tests). Moreover, ML has also been shown to deliver diagnostic visions when inspecting Electronic Health Records (EHRs). ML offer diagnoses based on multiple images, like computerized tomography (CT), magnetic resonance imaging (MRI), and diffusion tensor imaging (DTI) scans with 99.5% accuracy [23]. The following programming tools are available:

5.2.3 Python

Python is placed on the top most rank of all AI development languages due to its simplicity and ease to implement. Python is free and open source software which is freely available and distributable, even for commercial use [24]. Python is a programming language that entails a large number of standard libraries to process the clinical data. The extensively used libraries are Sklearn, Tensor Flow, Keras, PyTorch etc. A basic required library for a machine-learning classification task is represented in Fig. 5.1.

Various texts have been reported where a Python-based workflow has been developed for the diagnosis and prediction of cancer [25], heart disease [26,27], diabetes [18,28], etc. Python is also used to generate a deep neural network (DNN) by using Pytorch and Scikit-Learn to forecast death dates for patients with incurable illnesses.

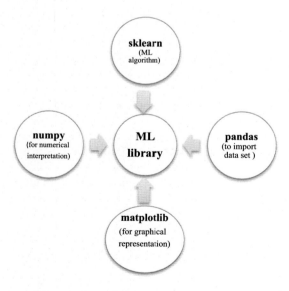

FIGURE 5.1

Basic library used in Python for ML.

5.2.4 MATLAB

MATLAB is also an ideal platform for applying machine-learning operations. MATLAB is a proprietary and closed software environment for development, while Python is an open source platform. MATLAB offers matrix algebra and a large network of data processing and plotting, along with the capability of ML. Statistics and Machine Learning Toolbox offer the capability of supervised and unsupervised ML algorithms such as K-NN, SVM, DT, k-means, k-medoids, clustering, Gaussian mixture models (GMM), and hidden Markov models (HMM). Medical images are collected for training and testing on the MATLAB image processing toolbox in order to diagnose skin disease [29] and tuberculosis [30].

5.3 Types of machine-learning algorithm

ML is a technique that offers computers the capability to learn without being explicitly programmed. Several real-life problems can be demonstrated by ML techniques. There are various types of ML techniques that are utilized in healthcare data for disease diagnosis system: (1) supervised learning; (2) unsupervised learning; (3) semisupervised learning; (4) reinforcement learning; and (5) deep learning, as displayed in Fig. 5.2.

Supervised learning: in supervised learning data is already categorized with the correct output. It is also called learning with a teacher or supervisor. KNN, DTs, SVM, neural network, Bayesian network etc., are some od the best-known algorithms of supervised learning. Supervised learning has two variations, i.e., classification and regression. Classification is used to predict categorical or discrete class label outputs for an instance. For example, "Is this tumor cancerous?" "yes" or "no" is classification. The regression is used to predict continuous

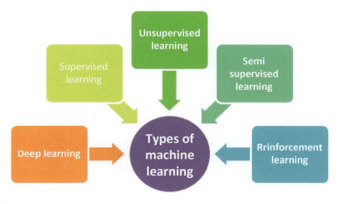

FIGURE 5.2

Types of ML.

quantity output for an instance. For example, it gives the answer of "How much" and "How many" nodes are infected or the area of spread of a disease.

Unsupervised learning: learning without a teacher is called unsupervised learning. This type of learning is completely based on experience and observation. Unsupervised ML algorithms are used to train a machine to determine hidden patterns from unlabeled data in the absence of target variables. The unsupervised machine learning is used for the identification of latent infectious diseases based on textual information of users about symptoms, temporal information in social media data, sentiment analysis, and body parts and pain location in a given location [31].

A Conditional Restricted Boltzmann Machine (CRBM) model provides a forecast of Alzheimer's disease evolution. A data set encompassing 18-month trajectories of 44 clinical variables of 1909 patients with Alzheimer's disease was used to train the system [32]. K-Mean, Self-Organized Model (SOM), and Expectation Model (EM) are some well-known algorithms of unsupervised learning.

Semisupervised learning: this type of learning associates both labeled and unlabeled data to create a suitable model or classifier. Self-training, generative models (GMM and HMM), S3VM, and graph-based algorithms are some popular algorithms of semisupervised learning [33]. Chai et al. (2018) suggested unlabeled gene expression instances for disease classification. An extended logistic regression model was intended that relies on semisupervised learning for disease classification [34].

Reinforcement learning: reinforcement learning is based on reward and punishment for the automation of any system. The system learns without the involvement of human and exploits its incentives and diminishes its penalties. Q-learning is the best-known reinforcement learning algorithm. WebMD and Mayo Clinic sites deployed an online symptom checker in order to recognize possible reasons and treatments for diseases based on the symptoms of patients. Symptoms were evaluated by asking multiple questions related to their symptoms and after that efforts are made to predict the potential diseases. Kao et al. (2018) suggested a context-aware symptom checking for disease diagnosis based on hierarchical reinforcement learning of a deep Q-network (DQN) [35]. Besson investigated several reinforcement learning algorithms and made them operable in a high-dimensional and reward-sparse setting for a rare type disease diagnostic system [36].

Deep learning: ML algorithms are suitable for those datasets that consist of around a hundred features or columns and formed as structured data. However, in the case of unstructured datasets, e.g., medical image, which entails a huge number of features, ML techniques become entirely unfeasible. For example a single 800×1000-pixel image in RGB color has 2.4 million features hence, conventional ML algorithms are inadequate to handle such a large number of features.

Deep learning is the extension of ML that is founded on the set of algorithms. In deep learning there are several layers of these algorithms and every layer offers a diverse interpretation of the data it feeds on. Such a type of algorithm is termed as ANN which is based on the interconnection of neurons and is analogous to a the neural networks present in the human brain. ANN, frequently called *deep learning*, permits computers to realize complex patterns in huge data sets. Deep learning is mainly appropriate for neurological disease [37]. CNN, DNN, and Recurrent Neural Network (RNN) are types of deep learning methods which have become dominant in numerous computer vision tasks, e.g., medical imaging [38], ultrasound image [39], and MRI [40].

5.4 Diagnosis model for disease prediction

Disease diagnosis is a serious and vital job performed by diverse intelligent systems that rely on computer technology. Any medical treatment begins with screening, diagnosis, treatment, and frequent monitoring. ML offers its capability to diagnose the disease at the initial stage. However, it is a very challenging task to recognize the disease at the initial stage. A common diagnostic model of various diseases is represented by Fig. 5.3.

With the clinical records and patient details available through different sources, the ML techniques can be utilized to create an evocative analysis of medical features. ML techniques are broadly applied in the diagnosis of numerous fatal diseases.

5.4.1 Data preprocessing

Data preprocessing is essential before its actual use. Data preprocessing is the concept of changing the raw data into a clean data set. The dataset is preprocessed in order to check missing values, noisy data, and other inconsistencies before executing it to the algorithm. Data must be in a format appropriate for ML. For example, if the algorithm processes only numeric data then if a class is labeled with "*malignant*" or "*benign*" then it must be replaced by "0" or "1." Data transformation and feature extraction are used to expand the performance of classifiers and hence a classification algorithm will be able to create a meaningful diagnosis. Only relevant features are selected and extracted for the particular disease. For example, a cancer patient may have diabetes, so it is essential to separate related features of cancer from diabetes. An unsupervised learning algorithm such as PCA is a familiar algorithm for feature extraction. Supervised learning is appropriate for classification and predictive modeling.

5.4.2 Training and testing data set

After the collection of relevant data, the next step is to divide the data set into a training set and a testing set. Most of the literature supports 70% of the data for the training set used for validating models and 30% of the data for testing the model to generalize the performance on unseen data. The dataset is split by using the function train_test_split (list of parameters) available in Python.

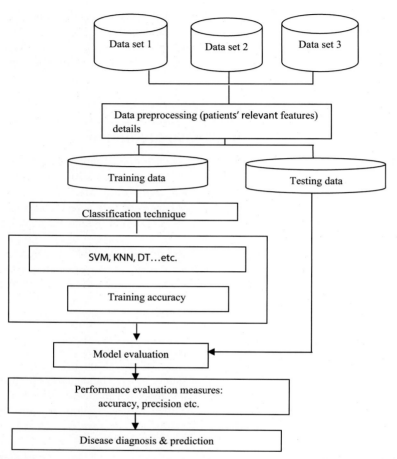

FIGURE 5.3

Common diagnosis models for disease.

The Train_test_split () function returns the list that consists of a train–test split of input and the corresponding target class, as shown in Eq. (5.1).

$$X\_train, X\_test, y\_train, y\_test = train\_test\_split \\ (X, y, test\_size = 0.3, random\_state = integer, shuffle = "true", stratify = "None") \quad (5.1)$$

5.4.3 Classification technique

The primary focus of the classification technique is to predict a category or class (y) from some inputs (X). Classification problems normally have a categorical output like a "disease" or "no disease," "1" or "0," "True" or "false." Some popular ML classification algorithms that are used for disease diagnosis are SVM, NB, DT, NB, KNN, and ANN. SVM and ANN are the most well-known ML algorithms in terms of diagnostic accuracy [1]. Some of the classification algorithms are discussed in Section 5.7.

5.4.4 **Performance metrics**

Different performance metrics are used to assess classification algorithms. These metrics are accuracy, sensitivity, specificity, and F1 score [41]. These metrics are calculated on the basis of the confusion matrix. Detail description of the confusion matrix is given in Section 5.6. After training and testing of the data set with different classification algorithms, the associated performance metrics are assessed to find the best suitable model. The selected model must have the highest value of performance metrics and can be deployed to diagnose the real-world samples.

5.5 **Confusion matrix**

A confusion matrix is a table that is used to define the performance of a classification algorithm. A confusion matrix visualizes and summarizes the performance of a classification algorithm. A confusion matrix is shown in Table 5.1, where *benign* tissue is called healthy and *malignant* tissue is considered cancerous.

The confusion matrix consists of four basic characteristics (numbers) that are used to define the measurement metrics of the classifier. These four numbers are:

1. TP (True Positive): TP represents the number of patients who have been properly classified to have malignant nodes, meaning they have the disease.
2. TN (True Negative): TN represents the number of correctly classified patients who are healthy.
3. FP (False Positive): FP represents the number of misclassified patients with the disease but actually they are healthy. FP is also known as a *Type I error*.
4. FN (False Negative): FN represents the number of patients misclassified as healthy but actually they are suffering from the disease. FN is also known as a *Type II error*.

Table 5.1 Confusion matrix.

	Predicted value	
Actual value	Malignant	Benign
Malignant	TP	FN
Benign	FP	TN

CHAPTER 5 Diagnosing of disease using machine learning

Performance metrics of an algorithm are accuracy, precision, recall, and F1 score, which are calculated on the basis of the above-stated TP, TN, FP, and FN.

Accuracy of an algorithm is represented as the ratio of correctly classified patients (TP + TN) to the total number of patients (TP + TN + FP + FN).

$$Accuracy = \frac{(TP + TN)}{(TP + FP + FN + TN)}$$

Precision of an algorithm is represented as the ratio of correctly classified patients with the disease (*TP*) to the total patients predicted to have the disease (*TP + FP*).

$$Precision = \frac{TP}{TP + FP}$$

Recall metric is defined as the ratio of correctly classified diseased patients (*TP*) divided by total number of patients who have actually the disease.

$$Recall = \frac{TP}{TP + FN}$$

The perception behind recall is how many patients have been classified as having the disease. Recall is also called as sensitivity.

F1 score is also known as the F Measure. The F1 score states the equilibrium between the precision and the recall.

$$F1Score = \frac{2*precision*recall}{precision + recall}$$

5.6 Disease diagnosis by various machine-learning algorithms

Health is wealth. Nothing is more important in this world other than health. Hence, there a need for appropriate disease diagnosis and prognosis systems. Various texts have been reported in the literature on various ML algorithms for disease prediction and diagnosis. Sophisticated algorithms must be trained with healthcare data before the system can help physicians with disease diagnosis and treatment plans. Supervised learning algorithms offer more clinically appropriate results compared to unsupervised learning in the healthcare system [1]. SVM, DT, Naïve Bayes, Random Forest, and KNN are some of the popular supervised ML algorithms used in this chapter. In this chapter, the author has taken three fatal diseases: cancer, neurological disease, and cardiovascular disease for analysis.

5.6.1 Support vector machine (SVM)

SVM is a supervised machine-learning algorithm that provides classification, prediction, regression, and outlier detection. SVM is a binary classifier. There are two categories in an x—y plane with points overlapping each other, and one has to find a straight line that can perfectly separate them and also maximizes the minimum margin. This maximum margin separator is determined by a subset of the data points that is often called "support vectors." The support vectors are specified by the circles around them, as represented in Fig. 5.4.

Functional margin of a point (x_i, y_i) w.r.t (w, b) is measured by the distance of a point (x_i, y_i) from the decision boundary (w, b)

$$\gamma_i = y_i(w^T x_i + b)$$

The larger functional margin means more confidence for correct prediction. For the optimal value, the function margin of (w.b) with respect to the training set should be the minimum of this functional margin:

$$\hat{\gamma}_i = \min_{i=1,\dots m} \hat{\gamma}_i$$

The functional margin offers a number, but without a reference it is tough to know the point which is actually far away or close to the decision plane. The geometric margin provides information not only as to whether the point is correctly classified or not, but also the magnitude of that distance in the form of units of |w|.

$$\hat{\gamma}_i = \frac{\hat{\gamma}_i}{w}$$

The geometric margin tells the distance between the given training example and the given plane. The SVM algorithm is realized in the form of a kernel. The kernel accepts a low-dimensional input space and converts it into a higher-dimensional space.

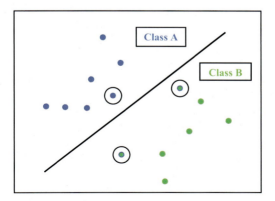

FIGURE 5.4

Classification using SVM.

100 CHAPTER 5 Diagnosing of disease using machine learning

After importing the SVM, the function SVC (*list of arguments*) with related parameters is available for use. In this chapter, we have passed the *'linear'* value of the parameter named *'kernel'* [38].

> *fromsklearn import svm*   import svm module from sklearn
> *sc = svm.SVC(kernel = 'linear')*   SVC function with Linear Kernel

The dataset is split by using Eq. (5.1). Then the model is fit on the training set by using *fit ()* function and a prediction is implemented on the test set by using the *predict ()* function.

> *sc.fit(X_train, y_train)*   Model is trained on train data set by SVM classifier
> *y_pred = clf.predict(X_test)*   Predict the response for test dataset

The performance of the model is estimated by how precisely the classifier can predict the disease of the patients. The accuracy of the model is represented by comparing the real value of the test set values and the model predicted values by the following function:

> *metrics.accuracy_score (y_test, y_pred)*   Accuracy of testing

5.6.2 K-nearest neighbors (KNN)

KNN is a nonparametric, supervised machine-learning method used for classification and regression. Predictions are made for a new input (X) by examining the complete training set for the K most like neighbors (instances) and summarizing the output variable for those K neighbors. Distance-based methods like *Euclidean* or *Manhattan* or *Minkowski* are used to define the K instances in the training dataset that are most similar in order to predict the target class label of the new input (X). Consider an uncategorized datum in Fig. 5.5 (shown as "?") and all other data is already categorized (star and diagonal), each one with your class (A or B). First calculate the distance from the new point to each point available in the data set. If $K = 3$ then take three of the smallest distance point class label and compare which ones have the smallest distance from the new data point or the target class label of the majority of points, for example in Fig. 5.5, the nearest data to the new data are those that are inside the first circle, and there are three different points and the majority of points belong to class B. Because there are two blue diagonal and only one brown stars, this new point is classified as blue diagonal or class B.

The KNN module has been imported *from sklearn.neighbors import KNeighborsClassifier* and the data set is split according to Eq. (5.1). The following parameters are defined in python for KNN classifier [42].

knn = KNeighborsClassifier(algorithm = "auto,"   leaf_size = 30,   metric = "minkowski,"   metric_params = None,   n_jobs = 1,   n_neighbors = 5,   p = 2, weights = "uniform")

Algorithm = "auto," selects the most suitable algorithm based on the arguments passed to *fit()* function. Algorithms like "ball_tree," "kd_tree," and "brute"

5.6 Disease diagnosis by various machine-learning algorithms 101

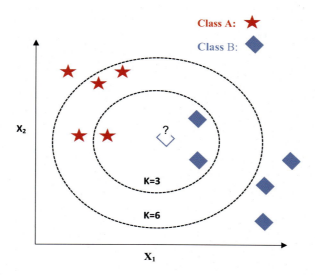

FIGURE 5.5

Classification by using KNN.

are used in KNN classifiers. The attribute leaf_size can affect the speed of the query building and essential memory status to store the tree. p is a power parameter for the *Minkowski* metric. If p = 2, then it is equal to the *Euclidean* metric and if p = 1, then it is known as the *Manhattan* distance. The metric_params is extra keyword parameters for the metric function. The *n_jobs* represents the number of similar tasks used to search the neighbor. The n_neighbors (k) is the number of neighbors for the queries. Weights = "uniform" represents that all points in each neighborhood have equal weight. Then the model is fit on the training set by using the *knn.fit(X_train, y_train)* method and the predict method is implemented on the test set by using the method as *y_pred = knn.predict(X_test)*.

The KNN algorithm has suitable in disease diagnosis systems such as heart disease prediction, in order to monitor the remote patient. KNN is used with a parameter weighting method to improve accuracy. A heart disease prediction system using KNN was suggested by [43] with basic patient health parameters where eight parameters are used from 13 parameters that can be instantly measured at home.

5.6.3 Decision tree (DT)

The DT is furthermost common supervised ML method that classifies ways to divide a data set in different situations. The objective is to build a model to predict the target class label by learning simple decision rules. The decision rules are primarily represented in the form of if—then—else statements. A decision tree has a structure like a tree where the node indicates "test" on an attribute (e.g., blood pressure is high or low), each branch indicates the outcome of the test, and a leaf node indicates the class label. Decision trees are actually made to understand and interpret the logic for the dataset. The decision tree is not like black box algorithms, such as SVM and NN.

Fig. 5.6 represents a DT for the prediction of heart disease. Chest pain (CP) and blood pressure (BP) are some potential features in order to diagnose patients suffering from heart disease. If CP is true, then disease is heart disease and if CP is false, then the next level of the decision node *BP* is tested for *high* and *low*. If the test result on BP node is high, then heart disease is *true*, otherwise *false*, which means the person is not suffering from heart disease. In order to implement the decision tree module in Python one has to import this module from Sklearn as:

From sklearn.tree import DecisionTreeClassifier

Use *DecisionTreeClassifier(list of parameter)* to train and test the data set for classification and prediction [42].

dt = DecisionTreeClassifier(criterion = "entropy," splitter = "best," max_depth = None, min_samples_split = 2, min_samples_leaf = 1, min_weight_fraction_leaf = 0.0, max_features = None, random_state = None, max_leaf_nodes = None, min_impurity_decrease = 0.0, min_impurity_split = None, class_weight = None, presort = False)

One of the possible values for the "criterion" is "entropy" and "Gini." Split the training and testing data set according to (5.1) and fit the classifier on the

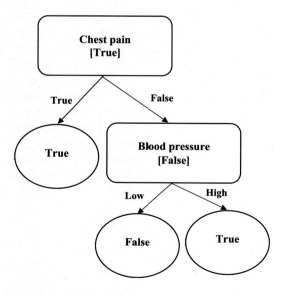

FIGURE 5.6

Decision tree classifier for heart disease.

5.6 Disease diagnosis by various machine-learning algorithms **103**

train attributes and labels by using *dt.fit(X_train, y_train)*. The trained classifier is used to estimate the labels of the test attributes as *dt.predict(X_test)*.

5.6.4 Naive bayes (NB)

The NB algorithm is suitable for datasets with a huge volume of features, but it is naive in the sense that it considers each feature as independent of one another. The NB algorithm is a collection of classification algorithms established on Bayes' theorem. Bayes theorem is shown in Eq. (5.2).

$$P(y|X) = \frac{P(X|y)P(y)}{P(X)} \tag{5.2}$$

where P(y | X) is a posterior probability, meaning the probability of target class y given the input feature X. P (X | y) is the probability of input feature X given that the target class y is true. P(y) is the probability of target class y. This is called the prior probability of y. P(X) is the probability of the input feature [44]. Naive Bayes is a supervised algorithm for binary and multiclass problems. Consider the Bayes theorem where $x_1, x_2, \ldots x_n$ are feature vectors and y is a target class label.

$$P(y|x_1, \ldots, x_n) = \frac{P(x_1, \ldots, x_n|y)P(y)}{P(x_1, \ldots, x_n)} \tag{5.3}$$

In a naive independence assumption, each feature is independent of one another, which reveals that the probability of a feature being stated relies only on the observation's class.

$$P(x_i|y, x_1, \ldots, x_n) = P(x_i|y)$$

Using the naïve assumption one can simplify Eq. (5.3) to consider the probabilities of features independently. Eq. 5.4 allows one to compute the probability that an observation belongs to a specific class for a given feature vector.

$$P(y|x_1, \ldots, x_n) = \frac{P(y) \sum_{i=1}^{n} P(x_i|y)}{P(x_1, \ldots, x_n)} \tag{5.4}$$

The denominator is always constant with respect to class; hence, one only needs to consider the numerator while determining the class of a given new input feature. The class of a new input feature is predicted as the maximum calculated probability of a given class, as given in Eq. (5.5).

$$y^{new} = \arg MaxP(y)\prod_{i=1}^{n} P(x^{new}_i|y) \tag{5.5}$$

In case of the Gaussian Naive Bayes model, a normal distribution model is fit for each feature of all of the possible classes. In python, in order to implement the naïve Bayes algorithm, first import the GaussianNB module *from*

104 **CHAPTER 5** Diagnosing of disease using machine learning

sklearn.naive_bayes import GaussianNB and by using *GausssianNB(parameter_list)* classifier, fit the algorithm on the training set and then predict the response for the test dataset.

5.7 ML algorithm in neurological, cardiovascular, and cancer disease diagnosis

5.7.1 Neurological disease diagnosis by machine learning

Neurological disorders are diseases related to the brain, spine, and the nerves that link them. There are approximately 600 diseases of the nervous system, such as brain tumors, epilepsy, Alzheimer's disease, Parkinson's disease, and stroke. Computer-aided and machine-learning diagnosis is critical to improve treatment policies for neurological diseases [45] due to inefficient, difficult, or worse diagnostic systems, that moreover are prone to human errors. Deep learning, an extension of machine learning, is able to extract the related specific features to different neurological diseases of the data set which are often in a large amount and high dimensions.

Magnetoencephalography (MEG) and electroencephalography (EEG) are imaging-based tools for the diagnosis of epilepsy, Parkinson's disease, and Alzheimer's disease [45]. The classification of MEG signals using deep learning minimizes the load of doctors and advances the accuracy of neurological diagnoses. Aoe [45] suggested a deep neural network technique, known as "MNet," to categorize neurological diseases and healthy from big data obtain from MEG signals. The trained "MNet" was performing well from SVM in terms of high accuracy and recall. The "MNet" exceeds the SVM in classification accuracy. Table 5.2 represents a proposed machine-learning algorithm used for the diagnosis of neurological disease from neuroimaging data. The treatment of neurological diseases can be achieved by reducing diagnostic error and increasing the accuracy of the algorithm. Table 5.2 also indicates the paradigm shift from machine learning to deep learning.

Table 5.2 Algorithms used in neurological disease.

S. No.	Neurological	Proposed algorithm
1.	Epilepsy	CNN [46], RBF & SOM [47], MNet [45], P-1D-CNN [48], Extreme Random Forest [49]
2.	Alzheimer	SVM with RBF [50], CRBM [32], multimodal deep learning [51], neural network [52]
3.	Parkinson's disease	Deep neural networks [53], deep neural network [54], Boosted Logistic Regression [55], Neural Network [56]

5.7.2 Cardiovascular disease diagnosis by machine learning

Cardiovascular disease is associated with the heart and blood vessels. When the heart is incapable of pushing the necessary volume of blood to other portions of the body system, it is called heart disease. Cardiovascular disease comprises heart attack, heart failure, stroke, aorta disease, abnormal heart rhythms, etc. The diagnosis of heart-related problems and their treatments are very complex and costly, particularly in developing countries due to a shortage of physicians, and the unavailability of diagnostic device and others resources, which may affect the accuracy of prediction and the treatment plan of the heart patients [57]. The eminent classifiers like logistic regression, SVM, KNN, DT, NB, ANN, and random forest are used to diagnose the heart disease based on the Cleveland heart disease data set [17].

A Decision Support System can be created to detect the heart based on a machine-learning technique that can have the best accuracy and performance relying on NB, SVM, logistic regression, random forest, and ANN etc. Numerous cardiovascular system parameters such as age, blood pressure, ECG values, gender, and blood sugar level are potential features to measure in order to predict the chances of the occurrence of heart disease.

5.7.3 Breast cancer diagnosis and prediction: a case study

Recently ML has been effectively implemented for cancer diagnosis and prognosis. The early diagnosis and prognosis of a cancer type is significant research to increase the survival rate of cancer patients. ML techniques can facilitate the subsequent clinical management of patients. ML methods have been used to recognize, detect, classify, or discriminate tumors and other malignancies. A variety of ML techniques, comprising ANNs, SVMs, NB, and DTs, have been extensively used in cancer research for the development of predictive models and accurate decision-making. Various types of cancer, such as breast, liver, lung, and cervical, are subjects of the study of ML algorithm. But in this chapter, we have particularly taken breast cancer as a case study.

Breast cancer is a heterogeneous tumor that has several subtypes with diverse biological behaviors, clinic pathological, and molecular characteristics. Currently there may be multiple reasons for breast cancer. These reasons are categorized into modifiable, nonmodifiable, and environmental. The modifiable reasons can be modified like menstrual and reproductive factors, radiation exposure, hormone replacement therapy, alcohol and high-fat diet, and modern lifestyle. Nonmodifiable reasons that cannot be changed by any person include age, gender, genetic factors, family history, and prior history of breast cancer. The environmental reasons are related to organochlorine exposure, electromagnetic field, and smoking [58].

106 CHAPTER 5 Diagnosing of disease using machine learning

5.7.3.1 Performance evaluation of breast cancer data set

Breast_cancer.csv data set used in this experiment is freely available as a dataset in Sklearn library [42]. That data set consists of 569 samples with 30 input features.

There are two target classes encoded as "0" for "*malignant*" and "1" for "*benign.*" In total 212 instances are *malignant* and 357 sample instances are "*benign.*" The data set is loaded and split as per Eq. (5.1) with test_size = 0.3 and random_state = 42.

After that we have applied various supervised classification algorithms and computed the performance metrics as represented in Table 5.3.

From the above Table 5.3, SVM offers better accuracy than other algorithms. The Accuracy of the SVM classifier is 96.49%, Precision is 96.36%, Recall is 98.15%, and F measure is 97.25%, thus it performs well for the diagnosis of breast cancer.

5.7.4 Impact of machine learning in the healthcare industry

Today, AI, ML, and deep learning are affecting each and every domain and the health industry is not an exception. AI and ML techniques are playing vital roles in the healthcare industry. AI and ML techniques are applied in six healthcare divisions, such as hospitals, pharmaceuticals, diagnostics, medical equipment and supplies, medical insurance, and telemedicine, to offer "better" healthcare services. ML will affect physicians and hospitals and play a significant role in clinical decision support systems. ML techniques will be able to detect disease at an early stage and also to tailor treatment plans to increase survival rate. Supervised and unsupervised learning algorithms have made major impacts in the detection and diagnosis of disease in healthcare. ML can impact on healthcare systems by improving efficiency and reducing the cost of treatment. ML and AI provide potential for medical imaging, which helps radiologists in the most proficient and cost effective way to diagnose and treat various diseases. An AI-based Chatbots application can recognize speech to identify patterns in disease. Patient photos are

Table 5.3 Performance metrics of applied algorithm.

S.N	Algorithm	Function name	Accuracy (%)	Precision (%)	Recall (%)	F1 score (%)
1.	SVM	SVC(kernel = "linear")	96.49	96.36	98.15	97.25
2.	KNN	KNeighborsClassifier (n_neighbors = 5)	95.91	94.69	99.07	96.83
3.	NB	GaussianNB()	94.15	94.55	96.29	95.41
4.	DT	DecisionTreeClassifier (criterion = "entropy")	95.91	96.33	97.22	96.77

being investigated using deep learning to discover phenotypes that are connected with rare genetic diseases. Deep learning is considered to be a powerful tool to examine CT, MRI, USG, mammography, ECG [59−61], and blood vessel detection [62]. IoT techniques can be used effectively to offer better health care to elderly people [63,64] and enhance quality when leveraged with machine learning, deep learning and Blockchain [65].

5.8 Conclusion and future scope

Machine learning has quickly adapted and transformed the healthcare sector. ML can help with the timely care of patients, reduced future risk of disease, and streamlined work processes. A conceptual diagnostic model is presented in this chapter, which is practically leveraged with Python. Supervised machine-learning algorithms are extensively used for disease diagnosis. Theoretical and practical concepts of SVM, KNN, DT, and NB have been introduced in this chapter. A confusion matrix is discussed to check the performance of a learning algorithm. Major disease areas where ML has proven its expertise are cancer, neurology, and cardiology. The SVM algorithm has been a better-suited algorithm for breast cancer, according to the case study. ML tools are very useful for disease diagnosis and also offer the opportunity to improve the decision-making process.

To conclude, it is believed that ML- and AI-based diagnostic system will offer smart healthcare applications in the coming years. Personalized medicine, automatic treatment, robotic surgery etc. are some of the future scope of AI and machine learning, which will be dominated by deep learning.

References

[1] F. Jiang, Y. Jiang, H. Zhi, Y. Dong, H. Li, S. Ma, Y. Wang, Artificial intelligence in healthcare: past, present and future, Stroke Vasc. Neurol. 2 (4) (2017) 230−243.

[2] H. Brody, Medical imaging, Nature. 502 (2013) S81.

[3] A.S. Ahuja, The impact of artificial intelligence in medicine on the future role of the physician, Peer J 7 (2019) e7702.

[4] K. Suzuki, Pixel-based machine learning in medical imaging, J. Biomed. Imaging 2012 (2012) 1.

[5] T. Kooi, G. Litjens, B. Van Ginneken, A. Gubern-Mérida, C.I. Sánchez, R. Mann, N. Karssemeijer, Large scale deep learning for computer aided detection of mammographic lesions, Med. image Anal. 35 (2017) 303−312.

[6] L. Zou, S. Yu, T. Meng, Z. Zhang, X. Liang, Y. Xie, A Technical Review of Convolutional Neural Network-Based Mammographic Breast Cancer Diagnosis, Computational Math. Methods Med. (2019) 2019.

[7] P. Singh, R. Agrawal, A Customer Centric Best Connected Channel Model for Heterogeneous and IoT Networks, J. Organ. End. User Comput. (JOEUC) 30 (4) (2018) 32−50.

[8] R. Fang, S. Pouyanfar, Y. Yang, S.C. Chen, S.S. Iyengar, Computational health informatics in the big data age: a survey, ACM Comput. Surv. (CSUR) 49 (1) (2016) 12.

[9] B. Ristevski, M. Chen, Big data analytics in medicine and healthcare, J. Integr. Bioinforma. 15 (2018) 3.

[10] S.B. Baker, W. Xiang, I. Atkinson, Internet of things for smart healthcare: Technologies, challenges, and opportunities, IEEE Access. 5 (2017) 26521−26544.

[11] X. Ma, Z. Wang, S. Zhou, H. Wen, Y. Zhang, Intelligent healthcare systems assisted by data analytics and mobile computing, Wirel. Commun. Mob. Comput. (2018) 2018.

[12] Dr. Amit Ray, https://amitray.com/what-holding-back-machine-learning-in-health-care/ (Accessed on 29 oct 2019).

[13] P. Sajda, Machine learning for detection and diagnosis of disease, Annu.Rev. Biomed. Eng. 8 (2006) 537−565.

[14] K.T. Chui, W. Alhalabi, S.S.H. Pang, P.O.D. Pablos, R.W. Liu, M. Zhao, Disease diagnosis in smart healthcare: innovation, technologies and applications, Sustainability 9 (12) (2017) 2309.

[15] G. Cho, J. Yim, Y. Choi, J. Ko, S.H. Lee, Review of machine learning algorithms for diagnosing mental illness, Psych. Invest. 16 (4) (2019) 262.

[16] WHO, Patient safety and risk management service delivery and safety, (2019). <https://www.who.int/features/factfiles/patient_safety/patient-safety-fact-file.pdf?ua = 1>

[17] A.U. Haq, J.P. Li, M.H. Memon, S. Nazir, R. Sun, A hybrid intelligent system framework for the prediction of heart disease using machine learning algorithms, Mob. Inf. Syst. (2018) 2018.

[18] V. Agrawal, P. Singh, S. Sneha, (2019). Hyperglycemia prediction using machine learning: a probabilistic approach. In International conference on advances in computing and data sciences (pp. 304−312). Springer, Singapore.

[19] E.A. El-Shafeiy, A.I. El-Desouky, S.M. Elghamrawy, (2018). Prediction of liver diseases based on machine learning technique for big data. In International conference on advanced machine learning technologies and applications (pp. 362−374). Springer, Cham.

[20] S. Sontakke, J. Lohokare, R. Dani, (2017). Diagnosis of liver diseases using machine learning. In 2017 International conference on emerging trends & innovation in ICT (ICEI) (pp. 129−133). IEEE.

[21] J.D. Mello-Román, J.C. Mello-Román, S. Gomez-Guerrero, M. García-Torres, Predictive models for the medical diagnosis of dengue: a case study in paraguay, Comput. Math. Methods Med (2019).

[22] M. Nilashi, H. Ahmadi, L. Shahmoradi, O. Ibrahim, E. Akbari, A predictive method for hepatitis disease diagnosis using ensembles of neuro-fuzzy technique, J. Infect. Public. Health 12 (1) (2019) 13−20.

[23] Bart Copeland, CEO, ActiveState, advancing opportunities in healthcare with python-based machine learning, (2019).

[24] P. Singh, R. Agrawal, Prospects of open source software for maximizing the user expectations in heterogeneous network, Int. J. Open. Source Softw. Process. (IJOSSP) 9 (3) (2018) 1−14.

[25] I. Mihaylov, M. Nisheva, D. Vassilev, Application of machine learning models for survival prognosis in breast cancer studies, Information 10 (3) (2019) 93.

[26] M. Padmanabhan, P. Yuan, G. Chada, H.V. Nguyen, Physician-friendly machine learning: a case study with cardiovascular disease risk prediction, J. Clin. Med. 8 (7) (2019) 1050.

[27] A. Acharya, (2017), Comparative study of machine learning algorithms for heart disease prediction.

[28] B.G. Choi, S.W. Rha, S.W. Kim, J.H. Kang, J.Y. Park, Y.K. Noh, Machine learning for the prediction of new-onset diabetes mellitus during 5-year follow-up in nondiabetic patients with cardiovascular risks, Yonsei Medical Journal 60 (2) (2019) 191−199.

[29] L.S. Wei, Q. Gan, T. Ji, Skin disease recognition method based on image color and texture features, Comput. Math. Methods Med (2018).

[30] V. Chandrika, C.S. Parvathi, P. Bhaskar, Diagnosis of tuberculosis using MATLAB based artificial neural network, IJIPA 3 (1) (2012) 37−42.

[31] S. Lim, C.S. Tucker, S. Kumara, An unsupervised machine learning model for discovering latent infectious diseases using social media data, J. Biomed. Inform. 66 (2017) 82−94.

[32] C.K. Fisher, A.M. Smith, J.R. Walsh, Machine learning for comprehensive forecasting of Alzheimer's Disease progression, Sci. Rep. 9 (1) (2019) 1−14.

[33] X. Zhu, (2007). Semi-supervised learning tutorial. In International conference on machine learning (ICML) (pp. 1−135).

[34] H. Chai, Y. Liang, S. Wang, H.W. Shen, A novel logistic regression model combining semi-supervised learning and active learning for disease classification, Scientific reports 8 (1) (2018) 13009.

[35] Kao, H.C., Tang, K.F., & Chang, E.Y. (2018, April). Context-aware symptom checking for disease diagnosis using hierarchical reinforcement learning.In Thirty-Second AAAI Conference on Artificial Intelligence.

[36] R. Besson, E.L. Pennec, S. Allassonnière, J. Stirnemann, E. Spaggiari, Neuraz, A. (2018). A model-based reinforcement learning approach for a rare disease diagnostic task. arXiv preprint arXiv: 1811.10112.

[37] A.A.A. Valliani, D. Ranti, E.K. Oermann, Deep learning and neurology: a systematic review, Neurol. Ther. (2019) 1−15.

[38] G. Litjens, T. Kooi, B.E. Bejnordi, A.A.A. Setio, F. Ciompi, M. Ghafoorian, et al., A survey on deep learning in medical image analysis, Med. Image Anal. 42 (2017) 60−88.

[39] E.M.A. Anas, P. Mousavi, P. Abolmaesumi, A deep learning approach for real time prostate segmentation in freehand ultrasound guided biopsy, Med. Image Anal. 48 (2018) 107−116.

[40] E. Gong, J.M. Pauly, M. Wintermark, G. Zaharchuk, Deep learning enables reduced gadolinium dose for contrast-enhanced brain, MRI J. MagnReson Imaging 48 (2018) 330−340.

[41] A. Tharwat, Classification assessment methods, Appl. Comput. Inform. (2018).

[42] F. Pedregosa, et al., Scikit-learn: machine learning in Python, JMLR 12 (2011) 2825−2830.

[43] I.K.A. Enriko, M. Suryanegara, D. Gunawan, Heart disease diagnosis system with k-nearest neighbors method using real clinical medical records, J. Telecommun. Electron. Comput. Eng. (JTEC) 8 (12) (2016) 59−65.

[44] K. Vembandasamy, R. Sasipriya, E. Deepa, Heart diseases detection using naive bayes algorithm, Int. J. Innovative Sci., Eng. Technol. 2 (2015) 441−444.

[45] J. Aoe, R. Fukuma, T. Yanagisawa, T. Harada, M. Tanaka, M. Kobayashi, H. Kishima, Automatic diagnosis of neurological diseases using MEG signals with a deep neural network, Sci. Rep. 9 (1) (2019) 5057.

[46] M. Zhou, C. Tian, R. Cao, B. Wang, Y. Niu, T. Hu, J. Xiang, Epileptic seizure detection based on EEG signals and CNN, Front. Neuroinformatics 12 (2018) 95.

[47] A.H. Osman, A.A. Alzahrani, New approach for automated epileptic disease diagnosis using an integrated self-organization map and radial basis function neural network algorithm, IEEE Access. 7 (2018) 4741−4747.

[48] I. Ullah, M. Hussain, H. Aboalsamh, An automated system for epilepsy detection using EEG brain signals based on deep learning approach, Expert. Syst. Appl. 107 (2018) 61−71.

[49] Y. Jerry, Epileptic seizure classification ML algorithms, binary classification machine learning algorithms in Python, (2019), <https://towardsdatascience.com/seizure-classification-d0bb92d19962>

[50] M. Grassi, G. Perna, D. Caldirola, K. Schruers, R. Duara, D.A. Loewenstein, A clinically-translatable machine learning algorithm for the prediction of alzheimer's disease conversion in individuals with mild and premild cognitive impairment, J. Alzheimer's Dis. 61 (4) (2018) 1555−1573.

[51] G. Lee, K. Nho, B. Kang, K.A. Sohn, D. Kim, Predicting Alzheimer's disease progression using multi-modal deep learning approach, Sci. Rep. 9 (1) (2019) 1952.

[52] J. Albright, Alzheimer's Disease Neuroimaging Initiative, Forecasting the progression of Alzheimer's disease using neural networks and a novel preprocessing algorithm, Alzheimer's Dementia: Transl. Res. Clin. Interventions 5 (2019) 483−491.

[53] T.J. Wroge, Y. Özkanca, C. Demiroglu, D. Si, D.C. Atkins, R.H. Ghomi. Parkinson's disease diagnosis using machine learning and voice. In *2018 IEEE signal processing in medicine and biology symposium (SPMB)* (2018) (pp. 1−7). IEEE.

[54] S. Grover, S. Bhartia, A. Yadav, K.R. Seeja, Predicting severity of Parkinson's disease using deep learning, Procedia Comput. Sci. 132 (2018) 1788−1794.

[55] C. Rubbert, C. Mathys, C. Jockwitz, C.J. Hartmann, S.B. Eickhoff, F. Hoffstaedter, M. Südmeyer, Machine-learning identifies Parkinson's disease patients based on resting-state between-network functional connectivity, Br. J. Radiology 92 (2019) 20180886.

[56] I. Tsoulos, G. Mitsi, A. Stavrakoudis, M.D. Papapetropoulos, Application of machine learning in a Parkinson's disease digital biomarker dataset using neural network construction (NNC) methodology discriminates patient motor status, Front. ICT 6 (2019) 10.

[57] S. Ghwanmeh, A. Mohammad, A. Al-Ibrahim, Innovative artificial neural networks-based decision support system for heart diseases diagnosis, J. Intell. Learn. Syst. Appl. 5 (3) (2013) 176−183.

[58] R.D. Nindrea, T. Aryandono, L. Lazuardi, I. Dwiprahasto, Diagnostic accuracy of different machine learning algorithms for breast cancer risk calculation: a meta-analysis, Asian Pac. J. Cancer Prevention: APJCP 19 (7) (2018) 1747.

[59] M. Jangra, S.K. Dhull, K.K. Singh, Recent trends in arrhythmia beat detection: a review, in: B.M. Prasad, K.K. Singh, N. Ruhil, K. Singh, R. O'Kennedy (Eds.), Commun. Comput. Syst., 1st ed., 2017, pp. 177−183.

[60] M. Jangra, S.K. Dhull, K.K. Singh, ECG arrhythmia classification using modified visual geometry group network (mVGGNet), J. Intell. Fuzzy Syst. (2020) 1−15 (Preprint).

[61] S.K. Dhull, K.K. Singh, ECG beat classifiers: a journey from ANN to DNN, Procedia Comput. Sci. 167 (2020) 747−759.

[62] S.S. Mondal, N. Mandal, A. Singh, K.K. Singh, Blood vessel detection from Retinal fundas images using GIFKCN classifier, Procedia Comput. Sci. 167 (2020) 2060−2069.

[63] R.S. Pandey, R. Upadhyay, M. Kumar, P. Singh, S. Shukla, (2020). IoT-based helpagesensor device for senior citizens. In International conference on innovative computing and communications (pp. 187−193). Springer, Singapore.

[64] J. Padikkaparambil, C. Ncube, K.K. Singh, A. Singh, (2020). Internet of things technologies for elderly health-care applications. In emergence of pharmaceutical industry growth with industrial IoT approach (pp. 217−243). Academic Press.

[65] P. Singh, N. Singh, Blockchain with IoT and AI: a review of agriculture and healthcare. Intl. J. Appl. Evol. Comput. (IJAEC), 11 (4) (2020) 13−27.

CHAPTER 6

A novel approach of telemedicine for managing fetal condition based on machine learning technology from IoT-based wearable medical device

Ashu Ashu and Shilpi Sharma
Department of Computer Science, Amity University, Noida, India

6.1 Introduction

Medical field is becoming a very large and extensive business. As described by researchers [1], the healthcare system has become patient-centered rather than disease-centered. And also healthcare delivery has become based on value more than that of volume. As big data has a huge volume of data, then it is important to extract value from this huge volume. The old data management methods are not satisfactory to analyze big data as different data sources generate data with vast variety and volume. Hence, we need the latest and inventive tools and technologies to handle big data that can further allow the ability to manage healthcare data.

The big data are allowing the prediction of diseases earlier than they appear on the basis of medical records. Before discussing everything in detail, we need to know what big data is.

6.2 Healthcare and big data

Fig. 6.1 shows the various sources that are responsible for generating data in huge amounts in the healthcare industry. Patient portals, search engines, health records, etc. are giving rise to big data related to health and associated fields. Big data analysis is capable of altering the technique of healthcare providers in order to raise attentiveness from the medical and other data sources.

Big data is indeed an evolving development which is characterized by several kinds of data sources that have huge volumes, great velocity, and extensive

114 CHAPTER 6 A novel approach of telemedicine for managing

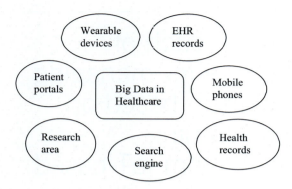

FIGURE 6.1

Sources responsible for big data in healthcare.

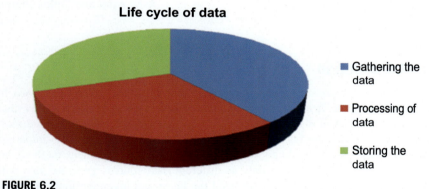

FIGURE 6.2

Life cycle of data [2].

variety of data. It allows the administrations to collect, store, handle, and examine massive amounts of data at the correct pace, at an appropriate time, to achieve the precise visions from difficult, noisy, diverse data.

The healthcare era uses analytics to convert data into some proper facts by generating some insights that lead to smarter business and clinical judgments, like dropping the number of readmissions of patients, recognizing and eliminating unwanted treatment, better clinician workflow etc.

Both government and private areas are concentrating on bringing in big data to provide better, faster, and additional appreciated services to individuals [2].

As per Fig. 6.2, there are three stages in the life cycle of data: gather, process, and store the data. Firstly, it is acquired through several methods and then the data is stored as per the user's liking. Then the acquired data is processed to generate some useful insights.

Fig. 6.3. Shows important 6 Vs that define big data

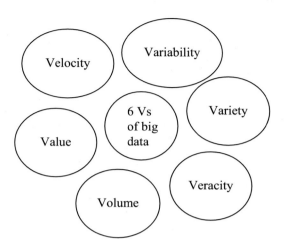

FIGURE 6.3

6 Vs of big data.

Variety rises from largely distinct sources or mashed-up data resulting from independent sources in different layouts. In healthcare, data can be observed to be organized, semiorganized, or unorganized. Laboratory data, medical, instrument data, and relational databases data come under organized. Data stored in XML format come under semiorganized, and free textual data that commonly do not have a specific design like manually written note data from X-ray images are unorganized.

Velocity corresponds to both the speed of data manufacture and the promptness of data management to encounter requests.

Veracity denotes the exactness and precision of facts. Big data may not be overall accurate, it is low in terms of veracity and therefore it is challenging to validate. As most of the data comes from new and unknown sources, it is important to set up a standard to assure the feature of the data.

Variability mentions data change ability during the handling of data. Developing range and inconsistency also raises the attraction of data and the chance of providing valuable information that is unexpected and unseen.

Value is the process of pulling out valued information from vast groups of facts. Value of data is beneficial for suitable decision-making.

6.3 Big data analytics

The data sets are huge and composite, challenging the recent practices to study and capture the conclusions. To make conclusions in the quickly growing business area, the big data analytics examines the data to reveal the unseen pattern,

116 CHAPTER 6 A novel approach of telemedicine for managing

	Tools Used	Limitations	When to use
Descriptive analytics *What happened and why?*	➤ Data Aggregation ➤ Data mining	➤ Snapshot of the past ➤ Limited ability to guide decisions	When you want to summarize results for all/part of your business
Predictive analytics *What might happen?*	➤ Statistical models ➤ Simulation	➤ Guess at the future ➤ Helps inform low complexity decisions	When you want to make an educated guess at likely results
Prescriptive analytics *What should we do?*	➤ Optimization models ➤ Heuristics	➤ Most effective where you have more control over what is being modeled	When you have important, complex or time-sensitive decisions to make

FIGURE 6.4

Big data analytics.

recognize unknown correlations, and understand the market movements, customer likes, and further useful business facts [3].

Fig. 6.4 Explains various analytics, the tools used to achieve it, and the limitations associated with them. The primary benefits that healthcare can achieve from big data analytics are:

- Timely discovery plus check of epidemics.
- Exact recognition and treatment of diseases which have little treatment success.
- Detection of new cures based on genomics and patient profiling.
- Stoppage of insurance and medical-claim fraud.
- Rise in cost-effectiveness of healthcare institutions.

6.4 Need of IOT in the healthcare industry

IOT stands for Internet of Things. With reference to [4], the main problem generally that patients face, particularly in remote places, is unavailability of clinicians and dealing with critical conditions. Today to eliminate this kind of problem there is a need to practice new technologies.

IoT devices are used for healthcare monitoring. They have the capability to make patients feel safe and fit, and also to improve methods that physicians use to deliver care.

IoT in healthcare also increases patient engagement plus satisfaction by letting them devote extra time to interactions with their doctors.

6.5 Healthcare uses machine learning

The power of machine learning (ML) lies in detecting disease and in sorting. Categorizing health data enables surgeons in speed up decision-making in the clinic. Google recently produced a machine-learning algorithm that detects malignant tumors in mammograms; also researchers in Stanford University are applying deep learning to identify skin cancer [5]. ML, in terms of healthcare, supports the analysis of thousands of various data points and suggests conclusions, offers on-time risk readings, allocates resources, and has many more other applications.

6.6 Need for machine learning

Machine learning is a branch of artificial intelligence (AI), it provides the system with the capability to learn on its own and enhance itself from the newer data without being openly programmed again and again by the programmer. Machine learning aims at developing a computer program that has the ability to gather the data and learn from it. The process of learning starts with observing some pattern in the data that helps in designing the model that suggests some predictions in the future data.

Flow of machine learning process

Fig. 6.5 define the five steps in machine learning process

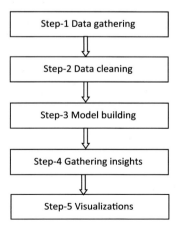

FIGURE 6.5

The machine learning process.

Data gathering is the first step, in which the suitable data is gathered for the model from dissimilar sources and loaded into the system's memory. It is significant to gather the interrelated data, as the accuracy and functioning of the algorithm based on the primarily gathered data.

Data transformation is the step of merging the data gathered from dissimilar sources. As soon as the data is merged then it is prepared or cleaned. Values that are null, missing, and duplicated are removed so that the algorithm works correctly.

Model building is the process of designing the algorithm with the help of the training data which yields higher accuracy.

Gaining insights is the process of collecting the results from the model.

Visualization is the process of showing the result with the help of graphs and bars that can be easy to understand.

6.7 Cardiotocography

Cardiotocography (CTG) in medicinal terms is a practical means of keeping track of the fetal heartbeat and uterine contractions during gestation, usually in the third trimester. This gives info on the uterine contractions, baseline fetal heart rate, and other evaluation parameters necessary for the obstetrician to check the well-being of the fetus and advise the expected time of delivery. Incorrect CTG values may suggest the possibility of facing delivery through Cesarean section. Therefore to guarantee precise prediction or a forecast of fetal well-being and also to make the mother ready for delivery, the CTG needs to be suitably inferred.

Until the 20th century, the evaluation of the baby situation in the womb was dependent on very few means: the growth of the uterus, the activities observed by the expecting lady, and checking the heart beat with a stethoscope. Unexpected nonappearance of baby activities in the second trimester of pregnancy was observed to be a serious problem. It was very important to record the condition of the fetus: if the fetal is alive or not, is it healthy or not? The continuing dilemma, whether the fetus was alive or expired in uterus, formed the main urge for the need of cardiotocography. As the heartbeat of fetus could be quite easily noticed by means of scans or through direct electrocardiography, cardiotocography became popular as the important method to monitor the fetal condition. It is the obstetrician's assurance that when the pattern of fetal heart rate (FHR) is observed to be normal there is nearly almost 100% conviction that the fetus is doing well. This made cardiotocography so eye-catching and has led to its extensive use. These days, cardiotocography is a well-known technique for monitoring the fetal well-being during pregnancy. Three skilled obstetricians created a UCI machine learning repository CTG data to categorize or classify the embryo into normal or suspect or pathological states. Normal, suspect, and pathologic are the classes in which data is classified.

6.8 Literature review

6.8.1 Research on revolutionary effect of telemedicine and its history

According to [6], using tele-monitoring and health informatics in a tele-health program, Veterans Health Administration (VHA) observed a 25% drop in records of bed days of care and a 19% reduction in number of hospital admissions as compared to usual care out of 17,025 participating patients. Starting from the beginning according to [7], "telemedicine" became a medical subject in 1993. It was earlier named as "consultation via television" and "diagnosis by television." The discussion was in both contexts of clinical and nonclinical. The first application was done in 1950. NASA also used telemedicine to check with the conditions of astronauts remotely using radio communication. Somewhere between 1960 and 1970, the US federal government also started to provide funding for seven telemedicine research and demonstration projects. With reference to [8], in 1970, telemedicine was applied to rural areas, remote areas, and distant areas. It worked well in Africa. By 1980 it had gained importance and visibility. Telemedicine turned out to lower the cost, traveling time, and hassles. Some medical sites that help in telemedicine are webMD.com, Medlineplus.gov, Medscape.com, and Mentalhelp.net. From the start the telephone was used, then television, then video conferencing, and so on. From wired to nonwired communication, some electronic devices that help in telemedicine are email services, interactive television, remote monitoring systems. And now devices are wearable and come with smart alarms. These alarms give information about an emergency and keep the patient under continuous monitoring by the specialist. Telemedicine is the ideal way of providing services to patient where and when they want it. In reference [9], the author talks to many doctors and asks about the changes happening to the health-care system using IOT. It discussed the several questions related to the benefits, cons, and misuses done by the patients. Doctors accept that it allows them to remotely monitor the patient's chronic medical conditions and provide care to them that is cost-effective and hassle-free too. Taylor [10] divided the telemedicine services into three broad categories, namely treatment, diagnostic, and educational

Ebad [11] discussed the different telemedicine applications in developing countries. For example, India has taken telemedicine initiatives and some of the departments use telemedicine.

The Department of Information Technology (DIT) initiated the projects for the full-scale standardization of telemedicine across the country. It has connected to 100 nodes all over India with the state governments. Indian Space Research Organization (ISRO) developed the Indian Satellite System (INSAT) for use in the implementation of telemedicine services across the country. It has developed its network with 382 hospitals, 306 in rural/remote areas. The initiative in rural area is taken by implementing GRAMSAT (rural satellite), for the development

120 CHAPTER 6 A novel approach of telemedicine for managing

of rural population. AIIMS New Delhi is also connected with rural areas of the states of J&K, Haryana, Orissa, Chandigarh, Punjab, and Himachal Pradesh over a network.

Apollo Hospitals are very well-known large private healthcare center. They also have developed some remote telemedicine centers. These centers connect villagers to specialists via satellites. Sood [12] stated that Aragonda was the first project carried for a rural telemedicine station developed by Apollo. This project helped villagers to connect to the specialists of Apollo hospitals located in Chennai. Now the centers are expanding to Bangladesh and other sites of the country. Apollo Hospitals are also providing facilities to the Army hospitals in South India via network hubs. Baater et al. [13] discussed the Mongolia Project that started in September 2007, and continued to December 2010. Its aim was to handle high-risk pregnancy of women in remote or rural areas. It included the consultations, prenatal scans diagnostics, fetal monitoring, and looked for other abnormalities. In total 297 health care workers were trained for this project in 2009 (March to December). 598 cases were referred, out of which around 383 were obstetrical, 127 were gynecological, and 88 were neonatal. Only 36 cases were referred for further diagnosis. Preston et al. [14] conclude that besides providing services to the patients, telemedicine system is also responsible for providing medical education and support to the rural health care workers. There are two modes of transmission existing, analog and digital. The digital mode transmits data in the form of a digital bit 0 and 1. Advantages of digital communications are their lower cost, simplicity, no need of heavy and costly equipment, and easy interfacing. MedNet is a telemedicine project. It provided the services of clinical consultations. It also allowed exchange of high-resolution X-ray images and other images between hospitals from rural areas and campus areas. Clinicians in the rural area hospitals also found it possible to share knowledge and can use fax machines. Around 90% of the fetal monitoring scans sent by fax machines could be managed on telephones. Hence MedNet offers many benefits to rural health care workers. Similarly [15], emphasizes the two situations when telemedicine should be considered. One is where there is no other option like in an emergency in remote surroundings. And another one is when it is better than the existing traditional services like teleradiology for rural health care centers. Pacis et al. [16] sum up the applications of telemedicine using artificial intelligence as monitoring of patient, healthcare information technology, intelligent assistance, and diagnosis and information analysis.

6.8.2 Role of machine learning in telemedicine/healthcare

Using machine learning algorithms, models are trained and developed with the training samples. These models help doctors and patients to detect early the diseases and keep monitoring them. Machine learning can still bring a revolutionary change in "Telemedicine." Hence the main focus is on discussing the growth of telemedicine from the "emergence of electronic devices to robots or models."

Using machine learning we can train our model with past experiences and it would predict results for new sample or instance. This model can be used to predict results of images like X-ray and ultrasounds. There are so many algorithms, it is important to check with which algorithm we are getting the best performance. Kumar and Kumar [17] performed a classification and developed a decision support system. Work was done with the help of technology, IOT, and machine learning. The goal was to recognize biosensor data in order to monitor and control the need of the patient. An SVM classifier was used for a small set. Parameters like systolic, diastolic, heart rate, body orientation were recorded. This one was one of the latest works done using machine learning algorithm. In 2017, an author [18] focused on patient monitoring, healthcare information technology, and intelligent assistance diagnosis and information analysis collaboration. It discusses different papers in terms of the services provided in the context of artificial intelligence, and draws attention to the implementation of artificial intelligence (machine learning), that can lead to cost-effective telemedicine.

In 2014, information analysis was done [19] using a KNN, data mining algorithms, to analyze behaviors of people with dementia. Also Guidi et al. [20] in 2014 suggested an HF clinical decision support system. This system has some sets of clinical parameters and enables support via tele-monitoring. This paper compares the accuracy of NN, SVM, random forest, and decision tree algorithms. The random forest algorithm gave the best performance in heart failure severity evaluation and prediction of heart failure type. In 2018, Duke and Thorpe [21] suggested an "Intelligent Diabetes Assistant" that improves diabetes. They worked on diabetes management techniques. They tracked the lifestyle, nutrition, and blood glucose readings of 10 diabetics using an intelligent diabetes assistant. It collected data, shared it with a physician, and automatically processed the data to extract important patterns. Data were collected through mobiles and armbands. Finally the data gave the idea of patient behavior. Methods were evaluated for identifying patterns and then to suggest changes for improved health. Ruiz [22] used medical interpretation for the early detection of anastomosis leakage. Feature selection is done by support vector machine linear maximum margin classifier. The result of early detection of complications after CRC was achieved to be sensitivity 100%, specificity 72%. These results can be used to develop prediction models. They support specialists and patients in taking decisions on time.

In 2018, Enriko et al. [23] used machine to machine technology with autodiagnosis features. Cardiovascular disease is a very dangerous disease. The author focused on rural areas where there are very few physicians. Patients from rural areas can use "My Kardio" which is a telemedicine system. This system has an autodiagnosing system that is developed with (KNN) machine learning algorithm. This architecture has three parts (patient site, server, and doctor site). After collecting the data, it was computed with KNN to give an autodiagnosis. The server site stores data on the database. The doctor checks the result and gives a recommendation. Akker et al. [24] were keen on activity monitoring and recording feedbacks. The aim was to determine a suitable classification algorithm and an optimal set of features that improve the performance of the classifier. In 2009, Fleury et al. [25] monitored daily activities and undertook a classification. This paper divided daily routing activities into seven points and the readings were collected through infra-red sensors, microphones, and wearable kinematic sensors. An SVM algorithm was applied for the classification of these activities. Experiment on 13 young individuals was performed to build a training set. Cross-validation was done to test this hypothesis. Ongvistepaiboon and Chan [26] developed a telerehabilitation system that helps patients to perform exercises at home. These is a smartphone-based system that handles frozen shoulder pain. For this smartphones with multiple sensors were required.

6.8.3 Role of big data analytics in healthcare

Ristevski and Chen [27] state that the healthcare industry produces—omics data like genomics, epigenomics, etc. and these raw data are useful for gaining insights toward various types of molecular profiles, changes, and interactions. These data are of different forms and produced by different sources.

Likewise EHRs data are also heterogeneous, can contain patient's personal data, diagnosis, charts etc. This huge amount of data should be collected, cleaned, stored, transformed, transferred, visualized, and delivered to clinicians in a suitable manner. Data mining techniques are used on EHRs data, web, and social media data in order to identify best practice guidelines for hospitals, identify association rules in EHRs data, and identify disease status. Guillen et al. [28] say that in general various systems do data acquisition using an IoT network. The key IoT modules are the sensors. The sensors are specialized medical peripherals. When we talk about a telemedicine model grounded on IoT architecture then there are necessary components like systems used to acquire data, gateways, computing structure, and healthcare benefactors.

Fig. 6.6, the data communication is among server and user devices. The user device sends messages continuously based on clinical parameters. The utmost common medical factors are heart rate, temperature, oxygen percentage saturation, etc. (Fig. 6.7).

Adibuzzaman and DeLaurentis [29] say that big data started with many convincing promises in the field of healthcare, but unluckily, medical science is dissimilar to other disciplines with added constraints like quality level of data, privacy, and some regulatory policies. The authors discussed these concepts in the hunt for a universal solution that allows data-driven conclusions to be decoded in health care, from bench to bedside. Existing big data structures are still at their early stages, and without handling these important matters the big data in healthcare may not attain its full potential. Hence to make it to the upper level, a larger group of institutions are needed to share additional complete, time-stamped, and precise data and with greater readiness to invest in expertise for deidentifying patient data to be worked on broadly for technical research. And also, as more and more "big data" structures are established, the technical and monitoring

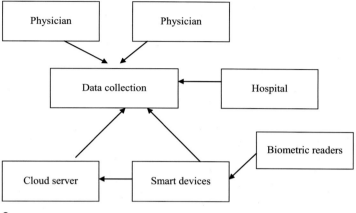

FIGURE 6.6

IOT network arrangement in telemedicine system.

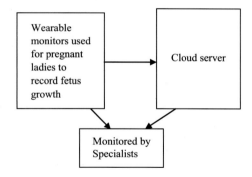

FIGURE 6.7

Process of collecting the data.

communities should be ready to think of new ways of catching relationships from data taken from routine healthcare (Fig. 6.8).

6.8.4 Challenges faced in handling big data in healthcare/ telemedicine

Undoubtedly telemedicine is a revolutionary change in health care. But with the advancement in the technologies, data are produced in tremendous amounts. And with the huge data, the main constraint is security. Reference [30] specially discussed about security. Security is also a determinant for the proper implementation of telemedicine. AAA comes under this security issue; standing for

LBE	LB	AC	FM	UC	ASTV	MSTV	ALTV	MLTV	DL	DS	DP	DR	Width	Min	Max	Nmax	Nzeros	Mode	Mean	Median	Variance	Tendency
120	120	0	0	0	73	0.5	43	2.4	0	0	0	0	64	62	126	2	0	120	137	121	73	1
132	132	4	0	4	17	2.1	0	10.4	2	0	0	0	130	68	198	6	1	141	136	140	12	0
133	133	2	0	5	16	2.1	0	13.4	2	0	0	0	130	68	198	5	1	141	135	138	13	0
134	134	2	0	6	16	2.4	0	23	2	0	0	0	117	53	170	11	0	137	134	137	13	1
132	132	4	0	5	16	2.4	0	19.9	0	0	0	0	117	53	170	9	0	137	136	138	11	1
134	134	1	0	10	26	5.9	0	0	9	0	2	0	150	50	200	5	3	76	107	107	170	0
134	134	1	0	9	29	6.3	0	0	6	0	2	0	150	50	200	6	3	71	107	106	215	0
122	122	0	0	0	83	0.5	6	15.6	0	0	0	0	68	62	130	0	0	122	122	123	3	1
122	122	0	0	1	84	0.5	5	13.6	0	0	0	0	68	62	130	0	0	122	122	123	3	1
122	122	0	0	3	86	0.3	6	10.6	0	0	0	0	68	62	130	1	0	122	122	123	1	1
151	151	0	0	1	64	1.9	9	27.6	1	0	0	0	130	56	186	2	0	150	148	151	9	1
150	150	0	0	1	64	2	8	29.5	1	0	0	0	130	56	186	5	0	150	148	151	10	1
131	131	4	57	6	28	1.4	0	12.9	2	0	0	0	66	88	154	5	0	135	134	137	7	1
131	131	6	147	4	28	1.5	0	5.4	1	0	0	0	87	71	158	2	0	141	137	141	10	1

FIGURE 6.8

UCI medical repository CTG data (raw data) [39].

authorization, authentication, and accounting. Privacy and information security are very much needed in the telemedicine system. Also it should be reliable as well. In telemedicine systems, complete and accurate information are needed to provide a successful service. When there is a security concern, that ultimately lowers the quality of care. Earlier security issues were not much highlighted but with the time it's important to take it into account. Researcher [9] discussed ethical violations because of the rapid increase in the usage of internet. It says ethical problem occurs in the stage of protection of privacy. Personal information is needed to be stored in a database that cannot be available to another person. There should be some rules and regulations. They also add that there is very little research published related to security, whereas security is the main constraint that makes telemedicine processes successful. SCT [31] claims that although telemedicine is a great revolution in healthcare, still it could not make any remarkable contribution to urban or rural areas. This paper takes an example of a girl who was in a bad condition. Doctors from urban place referred her to the hospital in Singapore. But the delay cost her her life. The authors said that her life could have been saved, if the doctors were able to get more suggestions from more reputed and knowledge physicians. Hence cloud-based telemedicine could help all persons in need, regardless of the location. It has been tried to resolve security issues by providing cloud technology. Telemedicine applications deployed in the cloud are compatible with a VOIP protocol that reduces the cost. SCT provided the connectivity between rural patients and doctors with experts. A body sensor was attached to the patient's end. All information is saved in the cloud. Only authorized person can access the information (Fig. 6.9).

In 2018, a paper [32] gave some security protocols. These protocols offer data confidentiality, integrity, authenticity, and untraceability. These are capable of handling user mobility along with patient untraceablity. They provide low computational complexity.

Healthcare data is very sensitive, with privacy and integrity a main attribute. Thus, in healthcare, big data security is very crucial. To provide the best service, a doctor must have fast and secure access to a Patient's health history. Rao et al. [33]

LB	AC	FM	UC	DL	DS	DP	ASTV	MSTV	ALTV	MLTV	Width	Min	Max	Nmax	Nzeros	Mode	Mean	Median	Variance
120	0	0	0	0	0	0	73	0.5	43	2.4	64	62	126	2	0	120	137	121	73
132	0.00638	0	0.00638	0.00319	0	0	17	2.1	0	10.4	130	68	198	6	1	141	136	140	12
133	0.003322	0	0.008306	0.003322	0	0	16	2.1	0	13.4	130	68	198	5	1	141	135	138	13
134	0.002561	0	0.007682	0.002561	0	0	16	2.4	0	23	117	53	170	11	0	137	134	137	13
132	0.006515	0	0.008143	0	0	0	16	2.4	0	19.9	117	53	170	9	0	137	136	138	11
134	0.001049	0	0.010493	0.009444	0	0.002099	26	5.9	0	0	150	50	200	5	3	76	107	107	170
134	0.001403	0	0.012623	0.008415	0	0.002805	29	6.3	0	0	150	50	200	6	3	71	107	106	215
122	0	0	0	0	0	0	83	0.5	6	15.6	68	62	130	0	0	122	122	123	3
122	0	0	0.001517	0	0	0	84	0.5	5	13.6	68	62	130	0	0	122	122	123	3
122	0	0	0.002967	0	0	0	86	0.3	6	10.6	68	62	130	1	0	122	122	123	1
151	0	0	0.000834	0.000834	0	0	64	1.9	9	27.6	130	56	186	2	0	150	148	151	9
150	0	0	0.000983	0.000983	0	0	64	2	8	29.5	130	56	186	5	0	150	148	151	10

FIGURE 6.9

CSV files of dataset.

emphasize that privacy and security issues are a major reason for anxiety as data is assured by international guidelines. In hospitals the applications that are affected by big data are genomics, patient care etc. The aim of authors was to suggest several feasible security solutions to increase the capability of big data relating to the healthcare sector in an extremely controlled environment.

The reference [34] discusses the key challenges generally faced while using big data analytics in the healthcare domain. These are gathering, storing, searching, sharing, and analyzing the collected data. Organizing the data after pulling out from different layers and combining it is also a challenging task. Hence, minimizing the faults that happen in medical decision support and verification is a subject of concern. The quality of information at each and every step should be tested by using proper safety methods. Additionally, a reduction in the costs of insurance in healthcare can also be beneficial for the patients.

6.8.5 Research done on tracing the fetal well-being using telemedicine and machine learning algorithms

Cardiotocography during gestation, along with ultrasonic scans, biophysical profile evaluation of fetus, etc., assures the well-being of fetus and achieves an initial identification of any worse condition. The motive is to keep a timely check on fetal condition. The cardiotocographic recording in a pregnant woman at high risk is regularly needed, which is worrying for patients who are away from equipped health centers like in rural areas. Due to this, telemedicine has been used in cardiotocography to reduce the pregnant women's referral to clinics or hospitals. It ultimately reduces the long stay in hospitals for patients at high risk. With reference to [35], on analysis of the last few years, pregnant women with a high-risk pregnancy or low-risk pregnancy undergoing cardiotocographic monitoring and combining it with telemedicine have better patient prognosis. The authors also talk about an Italian system that is telecardiotocography for antepartum.

Comerta [36] compared five efficient machine learning techniques for a fetal heart rate classification task, and the network performances were evaluated on several performance metrics attained from a confusion matrix. As per the results, ANN gave more proficient results than other machine learning algorithms. The attained results of every classifier were compared considering some performance metrics. Ved [37] worked on two states only, normal or pathological. The performance of each individual algorithm and model strength can be analyzed by performing a 10-fold cross-validation.

The medicinal field deals with the life of a person and a small error can be fatal. Hence, detecting outliers is important. An outlier is an observation that can be observed to be statistically distant from the remaining data or observations. A paper [38] concluded that the outliers may cause deviation or may complement the clinician's knowledge in building vital decisions with regard to the fetal birth and therefore require extraordinary attention.

6.9 Methodology

Methodology starts with the data life cycle:

1. Data Collection

    The starting stage of the process is gathering the data from several repositories and sources. Here the UCI machine learning repository is the source of data. Also if the data is so big, it can be stored in Hadoop Distributed File System (HDFS). The collected data can be both unstructured and structured like medical analysis, medicinal purchases, patient history, reports, wearable devices, and so on. Here in this work, data of pregnant ladies related to fetus are being collected as CTG data from a repository that can also be recorded using wearable monitors. Hospital managements can store it on the cloud and use it further for prediction. We took this data and stored it in a csv format. It also used the IoT framework.

2. Data Preprocessing

    After collecting the data from several sources it is stored in a common storage like csv file. This step helps in cleansing of data, sorting it accordingly for further exploration. In this step, all the missing values, rows, columns, or data are removed. Preparation of the data plays an important part in carrying the process forward. As raw data is tough to handle and also may not be helpful in achieving the appropriate result.

3. Data Reduction and Transformation

    After staging and preparing the entering data, this step helps in processing of the data from the pipeline by dropping the least important data (or) columns. Transformation of data aims in transforming the data by applying various mathematical and compression algorithms. This step does not lose any data but helps in sorting, to process further easily. Several analysis and analytic functions can be executed on the transformed data. Machine learning method will be fruitful if the selection of feature is good that can lead to improved prediction model and improved performance of the model.

4. Data Analytics

This step plays the major role in analyses of the data by drawing conclusions. It interprets the data that can be used further for research, knowledge, and business decisions. Using these results, industry models can be enhanced; also the unknown correlations can be made.

5. Data Output

The numerical, analytical information are presented based on the inputs given. The generated reports are understood appropriately in several formats for the user to view, check, or print. The final output data is the significant information that directs and supports in achieving the goal. To store the received output data for any further usage, a data storage system is required. Likewise, it is essential to uphold security, integrity, and data access control throughout the life cycle.

Here the proposed scenario is a type of classification and prediction problem as it is based on 21 extracted features of CTG data. Here, R language is used with a random forest machine learning algorithm. At preprocessing stage, we recognized the element that has noteworthy significance in the model. Generally, several performance measures are well-thought-out for model assessment, which has made the model more precise. For prediction performance a matrix was developed for model evaluation. The measures are accuracy, sensitivity (recall), specificity etc. Here on the dataset, model is build using random forest. The steps followed are:

- reading the data set;
- dividing the data set into training and testing;
- cross-validation set; and
- building the model using random forest.

Here, in this work data is taken and it has 21 features. We are taking into consideration 21 features from 40 features. For classification and prediction of the well-being state of fetal during pregnancy, the machine learning algorithm random forest is used, and the model is developed using R language. In total 2126 data points have been collected for these below-mentioned 21 features. After classification, the data is classified among three factors (NSP-1, 2, 3).

6.9.1 Preprocessing and splitting of data

Data taken from UCI machine learning repository obtained in Excel. It has 2126 instances with 40 elements like file name, segfile, and date. Normal state is denoted by 1, suspected by 2, while pathological is denoted by 3.

NSP is the response. To make it a factor, as it has integer value of 1, 2, or 3, and it is a deciding factor for classification and prediction. Hence data$NSP < - as.factor(data$NSP)

128 CHAPTER 6 A novel approach of telemedicine for managing

Table (data$NSP), gave the following table.

This table gives the number of results holding each value individually. Here we can see, the majority of data refers to normal factor.

NSP factor	1 (normal)	2 (suspect)	3 (pathologic)
Result	1655	295	176

```
For partitioning the data among training and testing:
Set. Seed ()
 Partition <- sample (2, n row (data), replace = TRUE, prob = c (0.7, 0.3)
 Train_data < - data [partition == 1,]
 Test_data < -data [partition == 2,]
```

Here a random seed has been used to make this analysis repeatable. For performing training and testing of model, data has been distributed 70:30. To train the model 70% of data is used and for testing purpose 30% of data is used. After this partition we observe that we have 631 interpretations in test data and 1495 interpretations in training data.

Here random forest algorithm is been used for classifying and predicting. Using this algorithm, the result can be withdrawn from aggregating many decision trees.

Random Forest Algorithm

This algorithm is performed by combining the trees.

It can be used for classification or regression.

It shuns over fitting.

It can deal with huge number of elements.

It helps with feature selection based on importance.

```
After dividing NSP into factors, some libraries and packages are need to
be installed.
install.packages("randomForest")
library(randomForest)
#LibraryRequiredinstall.
packages("plyr")
install.packages("caret")
install.packages("e1071")
library(plyr)
library(caret)
library(e1071)

#linear                                                         regression
data_lm < -lm(formula = NSP ~ ., data = dataset1)
summary(data_lm)
```

Fig. 6.10. shows important features, using linear regression that can focus on important features. These stars show the level of their impacts. The features that don't have stars at the end are least or negligibly important. Means these features

6.9 Methodology

```
(Intercept)   2.632e-01    1.347e-01    1.953   0.050914   .
LB            1.451e-02    2.218e-03    6.540   7.68e-11   ***
AC            4.528e-01    3.702e+00    0.122   0.902655
FM            5.334e-02    1.912e-01    0.279   0.780308
UC           -2.390e+01    3.223e+00   -7.416   1.75e-13   ***
DL           -9.858e-01    5.303e+00   -0.186   0.852527
DS            7.462e+02    1.417e+02    5.266   1.53e-07   ***
DP            3.855e+02    2.413e+01   15.974   < 2e-16    ***
ASTV          8.546e-03    6.737e-04   12.685   < 2e-16    ***
MSTV          9.347e-04    1.634e-02    0.057   0.954394
ALTV          1.179e-02    6.244e-04   18.886   < 2e-16    ***
MLTV          6.302e-03    2.153e-03    2.927   0.003463   **
Width         3.707e-03    9.044e-04    4.099   4.31e-05   ***
Min           7.000e-03    1.255e-03    5.579   2.73e-08   ***
Max                  NA           NA       NA         NA
Nmax         -2.520e-03    4.359e-03   -0.578   0.563191
Nzeros        1.129e-02    1.289e-02    0.876   0.381025
Mode         -4.099e-03    1.532e-03   -2.676   0.007500   **
Mean         -3.358e-03    2.412e-03   -1.392   0.164032
Median       -1.009e-02    3.005e-03   -3.359   0.000796   ***
Variance      2.675e-03    4.659e-04    5.742   1.07e-08   ***
Tendency      8.425e-02    2.323e-02    3.626   0.000294   ***
```

FIGURE 6.10

Linear regression.

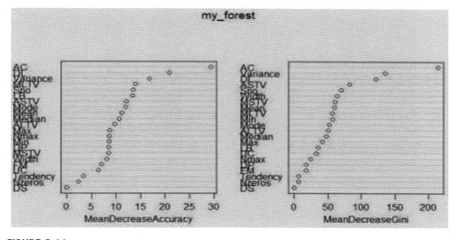

FIGURE 6.11

Crucial features.

can be eliminated or ignored. It is basic and is generally used in predictive analysis. The general use of regression is to test: (1) how much the set of predictor variables perform well in predicting a dependent result variable; (2) what variables are most important predictors to give result variable. Hence regression evaluations are very useful in explaining the association between one or more independent variables and one dependent variable.

Fig. 6.11 indicates the important features in prediction process and their impact level. Further random forest is implemented on these important features.

Fig. 6.12. shows confusion matrix and error rate for classification model of random forest. Here the number of trees is 500, and three variables are taken at

130 CHAPTER 6 A novel approach of telemedicine for managing

```
                 Type of random forest: classification
                        Number of trees: 500
No. of variables tried at each split: 3

        OOB estimate of  error rate: 6.76%
Confusion matrix:
        1    2    3 class. error
1 1143   13    4 0.01465517
2   64  147    2 0.30985915
3   12    6  104 0.14754098
```

FIGURE 6.12

Confusion matrix for classification model.

```
Confusion Matrix and Statistics

              Reference
Prediction    1    2    3
        1   486   21    5
        2     8   56    2
        3     1    5   47

Overall Statistics

                  Accuracy  :  0.9334
                    95% CI  :  (0.9111, 0.9516)
      No Information Rate   :  0.7845
      P-Value [Acc > NIR]   :  < 2e-16

                     Kappa  :  0.8058

   Mcnemar's Test P-Value   :  0.02053
```

FIGURE 6.13

Confusion matrix for prediction.

each split. The confusion matrix shows statistics for each class and each class error too.

```
#RandomForestImplimentation
rforest1 < -
randomForest
(NSP ~ LB + UC + DS + ASTV + ALTV + Width + Min + Median + Variance + Tendency,
data = train_data)
print(rforest1).

#Prediction
p1 < -predict(rforest1, test_data)
confusionMatrix(p1, test_data$NSP).
```

Fig. 6.13. gives the confusion matrix and other statistics. It gives an accuracy of 93.34%. Kappa score is 0.8058.

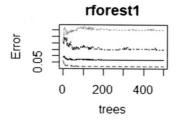

FIGURE 6.14

Error rate with respect to number of trees.

Fig. 6.14. gives the rate of error with respect to the number of trees. Using random forest algorithm we estimate **out of bag**, when we perform classification the out of bag error is basically accuracy. Thus at the tuning step the number of trees used is 300 and as can be seen from the figure, 200–300 is fine.

K-fold cross-validation can also be used here; it splits data into K subsections that reiterate the method for K times. Each time K-1 data is used to develop the model as training data and remaining is used for testing purpose. Computation time is less for a small dataset. Using K-fold cross-validation, the over-fitting problem can be handled. The CTG dataset is been split into a training and testing dataset for calculating the performance. This process is called K-fold cross-validation and avoids picking a part of data for training and testing.

6.10 Evaluation

Evaluating the performance

It is hard to guess performance for insufficient facts using machine learning methods. Hence, a minor volume of facts can be analyzed using the cross-validation method. It is important that the split of data into a training set and testing set must be completed prior to applying the machine learning approach. The whole CTG data have been divided into training_data and testing_data to evaluate the performance of the machine learning technique and later linear regression is done. The model designed using training_data is tested using testing_data in the process of classification. Efficacy of a classifier can be figured by number of true positive (TP), true negative (TN), false positive (FP), and false negative (FN) results. Sensitivity delivers positive test results, whereas specificity delivers negative test results ratio and complete amount shown by accuracy.

Accuracy, specificity, and sensitivity

Fig. 6.15. gives all the statistics. All parameters, sensitivity, specificity, accuracy, kappa score, etc., are given referring to all three classes, namely Class: Normal = 1, Class: Suspect = 2, and Class: Pathologic = 3. The prediction model also gives positively predicted values and negatively predictive values for each class.

```
                    Accuracy : 0.9334
                     95% CI : (0.9111, 0.9516)
       No Information Rate : 0.7845
       P-Value [Acc > NIR] : < 2e-16

                      Kappa : 0.8058

   Mcnemar's Test P-Value : 0.02053

Statistics by Class:

                        Class: 1  Class: 2  Class: 3
Sensitivity              0.9818   0.68293   0.87037
Specificity              0.8088   0.98179   0.98960
Pos Pred Value           0.9492   0.84848   0.88679
Neg Pred Value           0.9244   0.95398   0.98789
Prevalence               0.7845   0.12995   0.08558
Detection Rate           0.7702   0.08875   0.07448
Detection Prevalence     0.8114   0.10460   0.08399
Balanced Accuracy        0.8953   0.83236   0.92999
```

FIGURE 6.15

Statistics.

6.11 Conclusion and future work

This research uses a dataset from UCI CTG to estimate the machine learning classifier's performance and predictions using a random forest algorithm. This CTG dataset is classified into three states, namely normal, suspect, and pathological states. Future work will focus on measuring the performance of the classifier for a greater dataset. It is tough to find a biomedical dataset that can be trusted and well conserved due to confidentiality of patient data, unless big and reputed hospitals or health board associations initiate some data records for research. Fetal oddities can be identified by obstetricians with CTG statistics and medical operations can be applied before severe damage to the fetus. The number of trees and number of splits should be selected properly. The presence of outliers may diverge or accompany the obstetrician's experience in deciding better process or operations with regard to the fetus birth and therefore desire special attention. Also, we can use many outlier treatments. The performance of a model on a larger dataset is also our concern and will be taken care of in future work. The accuracy of the model should be enhanced further.

References

[1] S. Sa, B. Rai, A. Meshram, A. Gunasekaran, S. Chandrakumarmangalam, Big data in healthcare management: a review of literature, Am. J. Theoret. Appl. Bus 4 (2) (2018) 57–69.

References   **133**

[2] M. Singh, V. Bhatia, R. Bhatia, Big data analytics solution to healthcare, 2017 International conference on intelligent communication and computational techniques (ICCT).

[3] R. Sonnati, Improving healthcare using big data analytics, Int. J. Sci. Technol. Res. 6 (03) (2017).

[4] K. Dewangan, M. Mishra, Internet of things for healthcare: a review, Int. J. Adv. Manag. Technol. Eng. Serv. 8 (III) (2018).

[5] Flatworld Solutions Private Limited [IN] [online] Available on: <https://www.flatworldsolutions.com/healthcare/articles/top-10-applications-of-machine-learning-in-healthcare.php>.

[6] J. Kvedar, M. Coye, W. Everett, Connected health: a review of technologies and strategies to improve patient care with telemedicine and telehealth, Health Aff. 33 (2) (2014) 194−199.

[7] K. Zundel, AHIP: telemedicine, history, application and impact of librarianshipM.L. S. Bull Med Libr Assoc 84 (1) (1996) 71−78.

[8] A. Atac, E. Kurt, S.E. Yurdakul, An overview to ethical problems in telemedicine technology, Elsevier 13th international educational technology conference Oct 2013, pp. 116−121.

[9] Marry Ann Liebert, INC, Telemedicine: point/counterpoint, DOI HEAT- 2015-29006-chd, pp. 33−43.

[10] P. Taylor, A survey of research in telemedicine (1998) 63−70.

[11] R. Ebad, Telemedicine: current and future perspectives (2013) 242−248.

[12] S.P. Sood, India telemedicine venture seeks to improve care, increase access, Telemed. Today (2002) 23−26.

[13] T. Baatar, N. Suldsuren, S. Bayanbileg, K. Seded, Telemedicine support of maternal and new-born health to remote provinces of Mongolia, Stud Health Technol Inform 182 (2012) 27−35. Global Telehealth.

[14] J. Preston, F.W. Brown, B. Hartley, Using telemedicine to improve health care in distant areas (1992) 25−31.

[15] J. Craig, V. Patterson, Introduction to the practice of telemedicine (2005) 3−9.

[16] D.M.M. Pacis, E.D.C. Subido Jr., N.T. Bugtai, Trends in telemedicine utilizing, Artif. Intell (2017) 1−9.

[17] N.S. Kumar, P.N. Kumar, An intelligent decision − support system for telemedicine, 1 Sept 2018.

[18] Book: Integrating Telemedicine and Telehealth, 9-29.

[19] S. Zuai, S.I. McClean, C.D. Nugent, M.P. Donnelly, L. Galway, B.W. Scotney, et al., A predictive model for assistive technology adoption for people with dementia (2014) 375−383.

[20] G. Guidi, M.C. Pettenati, P. Melillo, A machine learning system improve heart failure patient assistance (2014) 1750−1756.

[21] D.L. Duke, Thorpe, Intelligent diabetes assistant: using machine learning to help manage diabetes (2008) 913−914. IEEE.

[22] C.S. Ruiz, Support vector feature selection for early detection of anastomosis leakage from bag-of-words in electronic health records, 5 September 2016, pp. 1404−1415.

[23] K.A. Enriko, M. Suryanegara, D. Gunawan, My kardio: a telemedicine system based on machine-to-machine (M2M) technology for cardiovascular patients in rural areas with auto-diagnosis feature using k-nearest neighbor algorithm (2018) 1775−1780.

[24] H. Akker, V. Jones, H. Hermens, Predicting feedback compliance in a teletreatment application (2010).

[25] A. Fleury, M. Vacher, N. Noury, SVM-based multi-model classification of activities of daily living in health smart homes: sensors, algorithm and first experimental results (2009) 274–283.

[26] K. Ongvistepaiboon, J.H. Chan, Smartphone-based tele-rehabilitation system for frozen shoulder using a machine learning approach (2015) 811–815.

[27] B. Ristevski, M. Chen, Big data analytics in medicine and healthcare, J. Integr. Bioinf. (2018) 20170030.

[28] E. Guillén, J. Sánchez, L. López, IoT protocol model on healthcare monitoring, Springer Nature Singapore Pte Ltd., 2017. Available from: http://doi.org/10.1007/978-981-10-4086-3_49.

[29] M. Adibuzzaman, P. DeLaurentis, J. Hill, B. Benneyworth, Big data in healthcare—the promises, challenges and opportunities from a research perspective: a case study with a model database, 1–9.

[30] V. Garg J. Brewer, Telemedicine security: a systematic review, 2011, 768–777.

[31] N. Jeyanthi R. Thandeeswaran, SCT: secured cloud based telemedicine.

[32] F. Rezaeibagha, Y. Mu Practical and secure telemedicine system for user mobility, 2017, 24–32.

[33] S. Rao, Suma & Sunitha, Security solutions for big data analytics in healthcare, 2015 second international conference on advances in computing and communication engineering, 2015 IEEE, 10.1109/ICACCE.2015.83.

[34] A.R. Reddy P.S. Kumar, Predictive big data analytics in healthcare, 2016 second international conference on computational intelligence & communication technology, 2016 IEEE, 10.1109/CICT.2016.129.

[35] A. Lieto, M. Campanile, M. Falco, I. Carbone, G. Magenes, M. Signorini et al., And prenatal telemedicine: a new system for conventional and computerized telecardiotocography and tele-ultrasonography, advances in telemedicine: applications in various medical disciplines and geographical regions.

[36] Z. Cömerta, A. Kocamaz, Comparison of machine learning techniques for fetal heart rate classification, ICCESEN 2016 132 (2017).

[37] M. Ved, Outlier detection and outlier treatment: a case study of the cardiotocography (CTG) dataset.

[38] Dataset: UCI medical repository. Available on: <https://archive.ics.uci.edu/ml/machine-learning-databases/00193/>.

[39] Big Data Analytics images [online] Available on: Gurobi optimization, <https://www.gurobi.com/company/about-gurobi/prescriptive-analytics/>.

CHAPTER

IoT-based healthcare delivery services to promote transparency and patient satisfaction in a corporate hospital

7

Ritam Dutta[1], Subhadip Chowdhury[2] and Krishna Kant Singh[3]

[1]*Surendra Institute of Engineering and Management, MAKAUT, Kolkata, India*
[2]*Durgapur Society of Management Science College, KNU, Asansol, India*
[3]*Faculty of Engineering & Technology, Jain (Deemed-to-be University), Bengaluru, India*

7.1 Introduction

In Indian currently, the healthcare delivery system is a primary object of concern. After 73 years of independence, India is trying to provide health care to all in every aspect. Now in India, as per newspaper reports, one allopathic doctor is available per 1457 population and all the AYUSH presented systems together have one doctor (allopathic, homeopathic, uniani, ayurvedic etc. together) per 868 population, which is better than the WHO recommendation of a 1:1000 ratio [1]. As per the IBEF report 2017, private hospitals have a 74% share of hospitals and 40% share of beds of the country [2]. The 2019 IBEF report revealed that about 50% of spending on in-patient beds is for lifestyle diseases due to rapid urbanization and modern-day living standards [3]. The healthcare sector is one of the largest employment sectors in 2019. As per 2017 Financial Year, the healthcare industry was the fourth largest employment sector employing 319,780 people [3]. Now, the healthcare sector is increasing in Tier-II and Tier-III cities. This scenario and 100% foreign direct investment (FDI) in wellness and medical tourism sector by AYUSH is creating more opportunity of players in this field.

Due to this expansion and the smart cities mission by Government of India [4], the smart telemedicine, AI, smartphone applications, and IoT-based services are in demand. These systems not only ease healthcare delivery but also increase the quality of care. And this effect reduces the chances for medical mishaps. The Indian healthcare sector is currently expanding in the medical tourism sector and smartphone-based healthcare providers of applications and pharmaceuticals. But this demand also raises a burning question on healthcare expenditure and transparency.

Machine Learning and the Internet of Medical Things in Healthcare. DOI: https://doi.org/10.1016/B978-0-12-821229-5.00001-X
Copyright © 2021 Elsevier Inc. All rights reserved.

135

Today the patients are well-educated and well-informed. Sometimes the patient and anxious relatives have their own version of medical care perception compared with the treating medical team. This increases the conflict between the healthcare service providers and consumers. Also, the proximity with competitors forced the sector to take innovative value additions to attract patients. Here, the market leaders want to take a cost-leadership position in the market. But on the other hand, the quality of the care can be measurable by various accreditations like NABH or JCI. To implement the quality of care in the healthcare sector and to maintain it has pushed the healthcare sector to invest finances, human resources, and time in the quality department. Here the innovative technology and coordination of the whole healthcare team plays an important role. In this discussion we discuss this factor elaborately and try to give a model of customer-centric services.

7.2 Uses of IoT in healthcare

Kevin Ashton first coined the term Internet of things in 1999. When several interrelating computing devices with a living and nonliving environment set up an ecosystem of interrelation, like transferring data without human—human or human—machine interaction with the use of unique identifiers (UID), it is called the Internet of Things. Others have provided similar definitions [5]. As per Kevin Ashton, this means using the internet to empower the computers to sense the real world by themselves [6].

When humans and machines are interrelated and exchange data and information to make a perfect ecosystem, it needs IoT in action. The healthcare applications related to Internet of things are as follows:

Some examples of IoT in healthcare services to date are: Ambient Assisted Living (AAL), The internet of m-Health thing (m-IoT), Adverse Drug Reaction (ADR), Community Healthcare (CH), Children Health Information (CHI), Wearable Device Access (WDA), Semantic Medical Access (SMA), Indirect Emergency Healthcare (EMH), Embedded Gateway Configuration (EGC), and Embedded Context Prediction (ECP) [7]. These systems are helping healthcare organizations to function smoothly and organize the future course of action in a real time frame.

AAL assists the elderly and specially-abled people to lead a healthy and risk-free life. Different sensors like motion sensors, cameras, and alarm-based systems designed to monitor the specially-able and elder people in a smart home basis are primary objective of AAL. It can remind about medicine, detect an accident or fall, alarm the nearest and dearest, or even call an ambulance on the spot.

m-IoT is a mobile-based sensor and communicating and computing technology which tracks different vital statistics of the patient through attached sensors and records on a real-time basis. It can easily detect any anomaly in the body of the patient and alarm the doctor or paramedics. Through this system, a patient can choose the dose of medication like COPD-dose or insulin shot. The real-time recording and monitoring system give the correct information like body temperature, glucose level, blood pressure, and ECG pulses [8,9]. *ADR* can be monitored through this system. It detects the adverse reaction or allergy, can indicate the lethal dose, and make a real-time alarm to the paramedics so the treatment can be started quickly, which can save the life of the patient.

CH and *CHI* are intelligent information and decision-making systems. Through ICT (information, communication and technology) it helps to build up telemedicine systems in remote areas where doctors are not available easily. Through the primary healthcare establishment, doctors sitting miles away can directly join the patient and discuss the problem, can collect the vital statistics of the patient through PHC, and prescribe a medicine or remedy to the needy. Healthcare kiosks are one of the examples of this system.

WDA is a part of m-IoT, where a patient can wear a device like a watch very easily and all the vital statistics of the patient's body are monitored in real time. Any anomalies can be detected on time and the primary treatments for saving life can be given based on the real-time data. It also maintains a database from where a patient can calculate his/her daily doses of medicine, whether they take it or not, timely alarms on the vital statistics, etc. The doctor can make a decision based on the past history of the patient provided by the clinically graded wearable devices.

SMA is an application where the patient gives authentication to the doctor to access the medical records or any other clinical data of the patient. This system ensures the security of the patient data, which are confidential, and make easy for the physicians to take care of the patients. This system maintains transparency and mutual trust and communication are increased.

EMH or indirect emergency medical care is an emergency situation-based smart system that can implement the IoT to access the accident location easily, and contact the nearest bloodbank, ambulance care system, and nearest hospital with accident and emergency department to plan mitigation of the situation in the shortest time in an effort to save life.

Patients are now connected to various medical equipments in Hospitals through *EGC*. The various data from a patient directly goes to the cloud-based system where the users like paramedics, nurses, or doctors use that through internet services and apply various medical equipment to intervene in the clinical situation. Automation and robotics has an important role to establish this. Similarly, *ECP* is used to build context-aware healthcare applications over the IoT network.

7.3 Main problem area of a corporate hospital

A corporate hospital has unique characteristics in India. The characteristics are as follows:

7.3.1 Location

The tertiary healthcare service providers are mostly situated in urban areas which are easily accessible by the patients, mostly nearby a bus stop, with a parking lot, and facilities for lodging and food in nearby areas. For this similarity, the trend is to set up tertiary healthcare facilities like multi- or super-speciality hospitals in close proximity to each other. This helps them to connect the patients easily but contradictorily a patient has more option in one go which increases the competition among hospitals. The share of the healthcare market depends on the value addition that similar hospitals may promote to the consumers.

7.3.2 Hassle on outpatient services

In India, almost 75% of the out-patient care in OPD is private, including the doctor's chamber and hospital OPDs [10]. The increasing lifestyle diseases in urban areas are increasing demand for speciality clinics like a diabetes center, oncology center, cardiac center of excellence, etc. Hospitals are have a tendency to establish speciality OPDs in their premises. Due to that, the booking of a doctor is a real-time problem for needy patients. Queue management is another problem in OPDs at peak hours. A shortage of waiting areas may increase the patient dissatisfaction.

7.3.3 Diagnostic services

Similarly, the pressure on diagnostic services like pathological laboratories or radiological imaging services is immense in a hospital at peak time. Proper queuing and maintaining the time are real-time problems for a hospital. Patient satisfaction depends on timely services, and the problem of this is the time mostly perceived by the patient him/herself as per the perception of cost and quality of the hospital. Here, the expectation and reality of services in hospitals and the patient's point of view may have some gap.

7.3.4 Inpatient services

The main problem faced by a corporate hospital is bed management. if the hospital has optimum bed occupancy of 85%–90% in some area, then the real problem is managing it. It is seen in major hospitals that patients are waiting to get a bed, and there is a long wait for formalities for the discharging of patients, specially TPA and corporate patients. In both ways dissatisfaction may emerge among patients.

7.3.5 Support and utility services

Hospital has to provide some support and utility services to run the operations smoothly. The various areas like canteen, housekeeping, pharmacy, basic supplies like water, toilets, newspapers in waiting hall etc., daily diet to inpatients, CSSD supplies, biomedical waste management system, linen and laundry, medical gas plants, equipment management, HVAC systems, transportation system in-situ and ex-situ, security, and other services need to coordinate with the hospital ecosystem to maintain the quality services to the consumers.

7.3.6 Coordination in medical section

With a high prevalence rate of noncommunicable lifestyle diseases in urban areas, most of the beds are reported to share specific patient categories. The improvement of medical care increases average life-year and life-expectancy rates. This increases the old-aged population who need special care and effort. Maintaining all the needs of a patient and the pressure of continuous monitoring about the health-related state of patients are a few problems related to clinical care. High bed-occupancy rate or in a few cases a very low occupancy rate due to bottleneck competition reduces the chances of an equal service to all patients and increases the chance of medical mistakes, accidents, hospital acquired infections, postoperative infections, low quality of the medical care, etc.

7.3.7 Medical record keeping

Electronic medical records are essential for all record keeping for the patient. This is a legal document too and the patient has to rely on the medical record for past medical history. The clinical audit and quality of care primarily depend on the therapeutic procedures which depend on timely accurate diagnosis and treatment processes. Through the medical record, one can easily understand the past history and treatment procedures of the patient. To maintain it correctly and to use it transparently, hospitals have huge opportunity through IoT-based systems and cloud computing.

7.3.8 Transparency

Throughout the whole medical procedures, the transparency over medical bills and pharmaceuticals are a major factor of patient satisfaction in India, where most of the inpatient medical bills are out-of-the-pocket [2]. Only 27% of the population is covered by health insurance [3]. Patients are educated and well-informed today and the medical bill is a major concern for choosing healthcare providers or for rethinking the choices. Through IoT-based information systems

140 CHAPTER 7 IoT-based healthcare delivery services

the consumer may get real-time feedback on the whole medical procedures and be well aware about the medical bill, which may get rid of some anxiety and worries.

7.3.9 Cost leadership model in market

In this highly competitive era, to retain the high-quality healthcare services the hospital must invest in quality equipment, quality procedures, and quality and highly trained professionals to maintain the supremacy in healthcare industry. On the other hand, they have to maintain the cost leadership in the market. Only through value addition and process reengineering can a hospital achieve this and IoT has biggest opportunity in this sector.

7.4 Implementation of IoT-based healthcare delivery services

When the healthcare delivery is the primary objective to achieve health for all in India, the private corporate industry is in tussle to capture the ever-growing market of tertiary medical care. India is becoming a destination of medical tourism with Rs. 195 billion market size in 2017 [3]. Online-based healthcare apps are increasing in the sector of online doctor consultation, doctors' bookings, and online medicine stores with home delivery.

Taking note of all of these problems and opportunities, the healthcare system needs a transformation to become a healthy ecosystem to form smart hospitals with modern facilities. Thus they could achieve quality of care at low cost to prevail in the market as a cost leader. The procedures are described as follows:

7.4.1 The work of value chain

The value chain model discussed by Michael E. Porter in his book competitive advantage published in 1985 [11], states in summary that value creation is adding values to a product or service to maintain a competitive advantage. Primary activities comprise inbound logistics, operations, outbound logistics, marketing and sales, and servicing. Secondary activities comprise procurement, human resource management, infrastructure, technological development, etc [11].

In today's competitive environment in the ever-growing healthcare sector, the application of competitive advantage is most important to maintain affordable quality healthcare with a maximum positive outcome in terms of healthcare delivery and profitability. Due to the problems faced by urban corporate hospitals all over the country, patients and consumers of the hospitals and healthcare providers are skeptical about their choices and this promotes dissatisfaction of the consumers. Many unlikely events of rejections and demolitions of public and private property and life-threatening events are examples of these [12—14].

Through the value chain model, these objections may be satisfied. To implement these, the idea of IoT and AI has huge opportunity in the field. We have a clear idea about ERP and MIS systems that were implemented in a hospital for smooth running the operations. But all of these are based on the activities on resources, logistics, and need human input. This sector can be evolved as a hybrid IoT environment inside a hospital when the IoT-based services through cloud computing and sensors are implemented to join the healthcare delivery model through which the patient's clinical condition and AI-based patient interfaces are connected successfully. It will shorten the time and involve minimum assets and interfaces.

Michael Porter and Elizabeth Olmsted Teisberg's book on "Redefining healthcare: Creating value-based competition on result" [15] discussed in detail the strategic involvement of value creation in a healthcare delivery scenario. The presence of IoT, AI, and various smart and hybrid tools to a hospital can turn it into a smart hospital where it can create value-based advantages to the hospital and for the patients. The main issue for reform of healthcare is [15] the relationships among the three factors: health insurance and access, structure of healthcare delivery, and standards for coverage.

As per Michael Porter, the points in the "Paradox of Healthcare" are:

1. Costs are high and rising.
2. Services are restricted and fall well short of recommended care.
3. In other services, there is overuse of care.
4. Standards of care lag and fail to follow accepted benchmarks.
5. Diagnosis errors are common.
6. Preventable treatment errors are common.
7. Huge quality and cost differences persist across providers.
8. Huge quality and cost differences persist across geographic areas.
9. Best practices are slow to spread.
10. Innovation is resisted.

These paradoxes clearly state that the competition does not working in these scenarios. Some zero-sum competitions are in the healthcare system, which do not create value to the patient regarding healthcare delivery.

To create this, Porter suggests the root cause is that all the competitions are in the wrong level and the wrong things. They are too broad as per between hospitals and healthcare providers, too narrow in performing services or interventions in niche area, and too local as they can help mostly the local community, thus the market share is based on a limited area [15]. The principles of value-based competition are:

1. The primary focus should be on value for patients, not just lowering the costs.
2. The competition must be unrestricted based on results.
3. Competition should center on medical conditions over the full care of the cycle [15].

To implement these, the IoT and AI have immense opportunity and credibility to redefine the healthcare as well as its strategic implementation, thus lowering the cost and creatingvalue for patients and value to the costs involved in the process.

The first step should be increasing the medical care facilities to diagnose and treat the patients correctly with low cost and high quality. The most important facilitator for a doctor to treat the patient is medical record of the patient.

7.4.1.1 IoT and medical record

The medical record system is now easily convertible into an e-medical record and it can be implemented into the present Hospital Information System. Here we can take the IoT as one of the tools to make an effective ecosystem of information provided to the doctors and nurses. Problem-oriented medical records are the most effective systems to diagnose and treat a patient effectively [16]. Here the patient is identified using an unique identification number called patient ID. The patient ID has direct linkup with the details of the patient and all doctor-related data that comprise the medical record usually. Here, we can add all the diagnoses and prescriptions of the patient with the medical record database using cloud computing (Figs. 7.1−7.4).

Through this system, the patient medical record will be stored in a systematic format and can be accessed from the patient information system by treating physicians. The medical history of the patients contains all the vital statistics, past medical history, and complications, all the diagnostic reports including all types of pathological and radiological diagnosis reports and recordings/photos, any disease prevalence like communicable or noncommunicable diseases, current medications, past surgical procedures, any history of known allergy, etc. All the data can help to form a retrospective study of the patient's health status which reduces the medical error, such as allergic reactions, duplicity of diagnosis or medicine, reaction between two or more different medications, detection of disease which has root cause in past or prevalence diseases, etc. It reduces the chance of wrong surgical procedures. As for example, a patient with a recent surgical history of cardiac arrest due to blockage cannot take medicines which are coagulator. A diabetic patient needs to reduce blood glucose level before any surgery, even if it is an emergency situation. Doctors can take effective decisions on a patient who is a regular at their hospital or provide the medical record as asked for any medical interventions. A clinical audit can easily use these methods, increasing the chance of quality treatment and reducing the error points. Today we need speciality treatment and all the department can check the patient data easily through a patient information system to coordinate easily. A patient with multiple problems like diabetes, urinary tract infection, and at the same time needing serious cardiac care due to a blockage needs a medical board to coordinate the treatment. Here, through this system the doctors are virtually present for a medical board and effectively achieve the plan to treat each complication as per the prescribed directions to not harm the patient adversely. This is the main system of a

7.4 Implementation of IoT-based healthcare delivery services

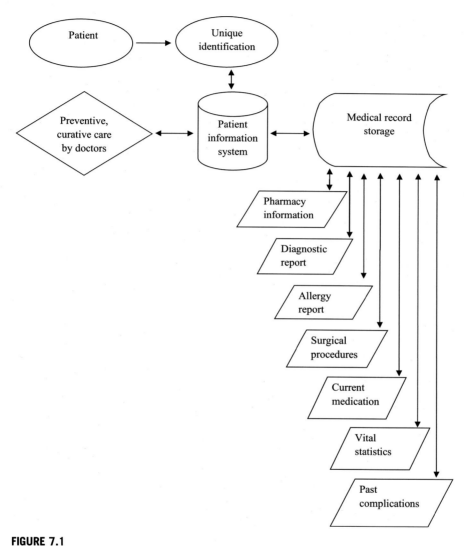

FIGURE 7.1

Use of electronic medical record storage for curative and preventive clinical decision-making through a patient's unique identification number and patient information system.

problem-oriented medical record and can be implemented easily through hybrid connectivity. For this the IoT is required. The IoT can create a system through cloud computing where the patient's medical record data are stored safely with low-cost virtual storage and are easily retrieved.

144 CHAPTER 7 IoT-based healthcare delivery services

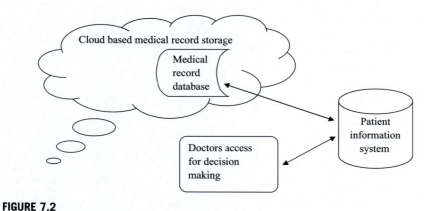

FIGURE 7.2

Cloud-based storage of electronic medical records.

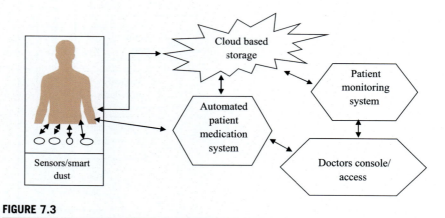

FIGURE 7.3

A 24 × 7 patient monitoring system and automated patient medication system controlled via doctor's console or direct automated intervention of emergency medicine and dose.

7.4.1.2 IoT and therapeutic facilities

After the effective utilization of the patient record, the next level should be the implementation part of the therapeutic facilities. The doctors can effectively make a treatment plan for the patient. To implement this plan, IoT has the most important role.

Effective diagnosis to implement IoT-based *Smart Dust* technology can be the answer to problems which are multidimensional. Through various magnetic resonance imaging technologies and radiographic technology we can make a 3D pictures of the patient viscera and using some pathological diagnosis the decision

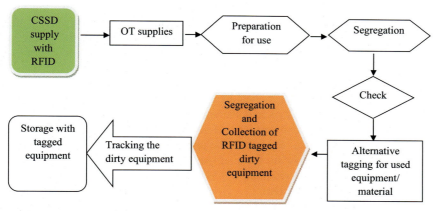

FIGURE 7.4

Tracking the clean and dirty supply.

can be made. But overuse of that technology can harm the patient in many areas like overexposure to radioactive radiation, piercing syringes multiple time or overuse of vein channels. Smart dust is tiny wireless microelectromechanical sensors (*MEMS*) which can detect the small changes in the environment. Once it is injected into the body, the sensors send all the data to the doctors to build a 3D image of the areas to detect the problem, easily take the information about live bloodstream, or dye the affected area. Through this procedure, only one single injection can diagnose the patient without many radiological exposure or syringes. The tiny particles easily detect the viruses or bacteria without stimulating them, and avoid the intelligent systems of those microorganisms which they used to hide themselves fofromrm antibodies. In the near future these smart dusts can directly treat the patient with the use of robotics and nanotechnology which reduce surgical interventions.

Apart from these, the IoT can be used for *monitoring* the patient effectively. The patients are nowadays surrounded and connected by various monitoring machines through which the patients are observed 24 hours a day. Any problem can alert the nurses to call the doctor. These monitoring systems can be modified through the use of IoT. Multiple sensors are attached to the patient's body and can send 24-hour data related to the patient directly to the cloud storage from where it can be accessed by the doctors. Any alarming situation can directly alert the doctor through the hospital information system or a smartphone-based application or pager without any. An *automated medication* facility is attached to the system which can easily give medications to the patient without disturbing the patient or without nurses' interventions. This automated medication system is effective for noncommunicable lifestyle diseases like diabetes where the insulin is released from the machine automatically as per the requirement of the doses. This

146 CHAPTER 7 IoT-based healthcare delivery services

type of intervention easily reduces the cost of the treatment because there is no need for frequent pathological tests or radiological tests to monitor the patient's clinical conditions. There is no need to disturb the patient and the medications can go automatically as prescribed by doctor or the doses can be adjusted as per the requirement of the time without changing the dose after the report. This will reduce the time for medical intervention. Through this, the error of treatment can be reduced and low-cost quality healthcare can be provided to the patient.

Apart from the diagnostics and therapeutic service, the IoT helps in multiple segments like *artificial limb* services. Here the artificial limbs can be connected and act as per neurological structure. The intelligence system can create the artificial connectivity and can be useful in amyotrophic lateral sclerosis (*ALS*) or motor neurone-related diseases particularly. Modern ergonomics can provide an effective way to maintain the system.

IoT can be used to monitor the patient's activity to reduce the *hospital-related injuries*. Continuous monitoring of the patient can therefore be avoided effectively. By *ambient-assisted living* the patient is tracked 24×7 by the IoT where it can detect a possible fall of the patient well in advance and call the housekeeping department to solve it. The non-CCTV tracking mechanisms in a bathroom can assist the patient to have access to easy toiletries, automated disinfectant liquids for toilet door handle and latrine assistant handles, automated flush and water supply, autospill proof floor, automatic alarm for patients who are busy in the bathroom more than 15 minutes for *prevention of fall or suicide*, etc. through *RFID* (Radio frequency identification) and *NFC* (near-field communications), which make the hybrid architecture of IoT by direct communicator (6LoWPAN) and passive communicator. These will protect the patient from hospital injuries and *nosocomial infections.*

Operation theater can be made safer by introducing smart IoT-based monitoring systems to detect any spill or cross-infections inside the room. The zoning of OT can be successfully implemented to use this system. The air-circulation, presence of infectious materials, the air-filter's work, the lighting, tracking of the biomedical waste collection, segregation and storage activities, identification of the clean and dirty path for clean sterile supply and biomedical waste collection through *RFID* tagging and *HVAC* (heating—ventilations—air-conditioning) control are the main area for IoT implementations in the whole OT operations and maintenance. The automatic controlled system easily detects and locks the processes in case of cross-infections.

7.4.1.3 IoT in supportive and utility services

The supportive and utility services associated with patient care are the second most important aspects of healthcare. To facilitate the clinical care to the patients effectively and efficiently, the supportive services' role is supreme. Through patient access management to inpatient services, it has widely accessed the area to promote the superior services.

7.4 Implementation of IoT-based healthcare delivery services   147

First the address covers the inpatient supportive services. To protect the patient, the controlling of access to the patient is primary; in case of an admitted patient the family members can access them once or twice a day for a particular visiting hour. Through IoT the traffic can be maintained systematically and can be scanned for potential cross-infectious agents. IoT-enabled services can scan the patient's relatives before their entrance to the inpatient areas. Then by an effective time management schedule they can enter into the patient area and visit the patient. For effective time management, the entry of the patient's relatives can be collected by the *RFID-enabled visitor's card.* Through this, the access can be controlled and the tracking of the patient's relatives inside the inpatient area can be restricted. The scanning or disinfection methods can be implemented per card per visit to control the cross-infections and detect the source.

The dirty *linen and laundry services* can be done through effective small sensors attached with each and every linen and other laundry services to track the dirty linen supply, its washing and cleaning facilities, supply of duly covered sterile linen supply, and the use of it in a patient's bed and related area (handkerchief, towel, hospital dress, etc.). By the IoT-enabled tracking system, the total supply of linen to the patient, average use of the linen supply per patient, dirty supply returned to the laundry and cleaner, and effective cleaning with sterile tagging for reuse are initiated in the integrated hospital information system to reduce the cross infection from hospital's side.

Inpatient diagnosis booking facilities enabled with *smart tools* can *track* the possible diagnosis order by the doctors and automate the prebooking of the diagnosis facility for the patient so the paramedics can take the patient for the diagnosis facility without intervening with the outpatient and emergency patient's diagnosis schedule. *Wearable device access* can cover the total time the patient takes for diagnosis and possible cross-infection or medical conditions for the procedure. Through IoT the waiting time and contact time can be calculated and the factors to better the time and limit the exposure can be pointed out.

Dietary services are being used in IoT based medical institutes for automated diet report generation of patients. In addtion to this, any lagging in nutrient intake can also be intimated by that smart system. Also, each dietary service can be packed with correct reference *bar-coding* to deliver it to the right patient at right time so that the nutrients are preserved in the food and the food is warm and tasty. It will limit the food adulteration problems associated with packing, timing, and delivering. The transportation of meals to the patient gives the signals to the other departments so that any medicine or diagnostic procedures before meals can be done timely without the error.

Odor controlling inside the inpatient area is a major emphasis in modern hospital facilities. Through IoT enabled sensors the hospital information system easily detects any abnormalities like fire alarm, smoke alarm, HVAC related issues, and infection warnings. The same system can be used to regulate the odor of the hospital, in particular area to make the right environment for the patient.

Tracking *biomedical waste* collection, segregation, transportation, and disposal procedures using IoT-enabled tracking facilities could control infection in the hospital.

7.4.1.4 IoT in patient delight

The patient-centric management system can create value to the patient by implementing various measures for patient support [17−20].

Through a *patient booking system*, the regular out patients flow to the hospital can be regulated systematically. The booking of the patient is directly accessed by the patient calling system, and can be managed by the emergency and age of the patient. The waiting time and queue can be shortened with this system. Patients, once they know about the timing of the doctor's visit, can control their emotions about the waiting time for doctor, queue management can be improved, and the hospital can maintain a minimum of chaotic moments.

Once the patient is booked for a doctor, the past history of the patient in case of followup patients and registered patients are automatically sent to the doctor's system.

A *bed management* system for optimum bed occupancy can be initiated through IoT-enabled services, where the total emergency patients, patients schedule for admission, and referral admission can be tracked effectively. Here, the doctors get alerts of the patient's present status and act accordingly. The patient's family members also get alerts about the movements which will reduce the anxiety of the patient's family.

TPA patient's insurance services providers get information about the patient's consumables so that the timing for procedures of insurance bills and discharge procedures of the patients shorten.

Bill management system: the patient-related dissatisfaction is generally due to the bill of the treatment. The present IoT-enabled system not only gives the information to the doctors and healthcare providers, it can have limited or restricted access for the patient relatives. The relatives, with the permission of the patient, can access the day-by-day bill live, information about day-to-day diagnosis and doctor visits, digital visiting cards, live information about a patient's change of bed or departments, alerts to any emergency situations, information about patient's diet and frequency, updates about bill, surgical procedures, payment facilities, discharge announcements, doctor's counseling time, or even book a doctor for counseling at their convenience etc. This system will facilitate transparent services to the patient and relatives to solve the confusion about billing and timing. Major bills for inpatients are out-of-pocket [10], only 27% are covered under insurance [3]. This system limits the dissatisfaction among Indian patients caused by private healthcare providers [12].

7.4.1.5 Cost leadership with quality of care

In today's competitive world, India is becoming a major hub for medical tourism [3]. For this Indian hospital require quality accreditations from quality standards.

In India, in total 21 hospitals are accredited as JCI (Joint Commission International) and most of the corporate hospitals have NABH (National Accreditation Board for Hospitals and Healthcare Providers). Through this they standardize their quality of care at international levels.

Implementation of quality standards costs initially at the introductory stages. Through the IoT it can be simplified and cost leadership can be formed. First, all the areas that are prescribed in the JCI or NABH or ISO standards are examined by an expert committee. Then the regular maintenance and observatory tools have to be initiated and IoT-enabled systems should be implemented with this to facilitate a hybrid ecosystem and closed-loop ecosystem with the services. The standard operating procedures should be automated to review the administrative and clinical system clearly. The patient information system should join all the supportive departments so that when the doctors add a prescription for diagnosis or medication, the pharmacy management system, radiology and pathology system are automatically notified about the prescription. Then they can systematically arrange for the same which is received by the nursing station automatically. This will maintain a systematic transportation system from pharmacy or pathology to the nursing station and vice versa without wasting valuable time using technology like a *bar-coded pneumatic transport system*. This will reduce the error and time drastically which reduces the cost and treatment error cost. All the reports should be uploaded to the patient information system automatically so the doctor can see and implement the measures. Doctors must have an online prescription system, where they can give digitally signed prescriptions and discharge summaries so that the billing and insurance department have time to release the patient as soon as possible. It will reduce the waiting time, cost of patient care, clinical error, and thus maintain the quality of the treatment with accuracy and save the valuable time and budget of the patient.

For this, the *administrative services* can join their hands to reduce the patient-related costs by using the IoT enabled database. The central store, medical store, and pharmaceutical stores can be jointly notified by the ERP system for requirements and lead-time on store situations live and can report to the financial management system for budget allocation and requirements online. Through this system, the online interventions from the HR manager, Operations Manager, CEO, General Manager, Finance Manager etc. are simplified. The decision-making procedures are simplified, and the patient care system becomes cost-effective in terms of outcome.

Accident and emergency system also benefitted by IoT implementation where the ambulance can easily track the traffic of the accident area and can choose the shortest route to the hospital. Then the emergency department can check the availability of blood directly from the blood bank to notify the patient's relatives about the additional requirements if any to limit the time of intervention.

Thus it creates value to the patients and creates a brand name in the healthcare industry. Urban hospital-related problems can be solved through this system.

7.5 Conclusion

IoT not only solved the urban hospital problems but can be initiated to everyday patient life to modify the lifestyle and reduce the lifestyle-related diseases. Different smartphone-based applications are available where patients can directly seek the help of doctors of a particular hospital to provide the unique identifications. The current status of the health can be monitored from home-based IoT programs. Smart hospitals limit the chance for accidents, limit the time for interventions, and minimize the cost related to quality of care. The IoT has a huge opportunity to build an AI-based framework for healthcare providers where the quality of treatment can be provided in a cost-effective manner, reducing the discrepancies and errors at a same time.

By introducing various wearable technologies, the smart hospital can manage the automated medication and intervention systems with life-supporting mechanisms which alert the system in adverse cases and manage the system without wasting time. From wheelchairs to artificial limbs, IoT can maintain its dignity in this information era. It can sound futuristic but the systems are already gaining attention and in use worldwide. The high cost of quality can be reduced through this system which can effectively be an antidote to the paradox of high-quality cost.

References

[1] PTI new Delhi report, 4th July, 2019, India has one doctor for every 1,457 citizens: Govt, Business Standards. <https://www.business-standard.com/article/pti-stories/india-has-one-doctor-for-every-1-457-citizens-govt-119070401127_1.html>, (accessed 10.12.2019).

[2] IBEF, Report on healthcare. <https://www.ibef.org/download/Healthcare-January-2017.pdf>, 2017 (accessed 10.12.2019).

[3] IBEF, Report on healthcare. <https://www.ibef.org/download/healthcare-jan-2019.pdf>, 2019 (accessed 11.12.2019).

[4] Government of India, smart cities list. <http://smartcities.gov.in/content/innerpage/list-of-projects.php>, smart city mission. (accessed 10.12.2019).

[5] Rouse, Margarate, June 2016, Internet of things. <https://internetofthingsagenda.techtarget.com/definition/Internet-of-Things-IoT>. (accessed 11.12.2019).

[6] T. Cole, 11th February 2018, Interview with Kevin Ashton-Inventor of IoT. <https://www.smart-industry.net/interview-with-iot-inventor-kevin-ashton-iot-is-driven-by-the-users/>. (accessed 12.12.2019).

[7] S.M.R. Islam, D. Kwak, M.H. Kabir, M. Hossain, K. Kwak, The Internet of things for health care: a comprehensive survey, published on 2015, *IEEE Access*, doi: 10.1109/Access.2015.2437951.

[8] M. Jangra, S.K. Dhull, K.K. Singh, ECG arrhythmia classification using modified visual geometry group network (mVGGNet), J. Intell. Fuzzy Syst. (2020) 1−15. Preprint.

[9] S.K. Dhull, K.K. Singh, ECG beat classifiers: a journey from ANN To DNN, Procedia Computer Sci. 167 (2020) 747−759.

[10] S. Ravi, R. Ahluwalia, S. Bergkvist, "Health and morbidity in India (2004−2014), Brookings India Research Paper No. 092016, 2016.

[11] M.E. Porter, Competitive advantage, New edition reprint 2004, Simon & Schuster, ISBN: 9780743260879, 0743260872.

[12] M. Krishnan, DW.com, Private hospital's exorbitant bill spotlights poor public healthcare in India, 22nd November, 2017. <https://www.dw.com/en/private-hospitals-exorbitant-bill-spotlights-poor-public-healthcare-in-india/a-41487085>. (accessed 11.12.2019).

[13] R. Arora, Economic times, 21st June, 2019, view: here's a treatment for mob violence against doctors in India. <https://economictimes.indiatimes.com/news/politics-and-nation/view-heres-a-treatment-for-mob-violence-against-doctors-in-india/articleshow/69889084.cms>. (accessed 12.12.2019).

[14] Mint, 31st August 2017, India's public health crisis: mint's reading list. <https://www.livemint.com/Opinion/Wl8s3ocQLKbuxOsESWNiVN/Indias-public-health-crisis-Mints-reading-list.html>. (accessed 11.12.2019).

[15] M.E. Porter, E.O. Teisberg, Redefining healthcare: creating value- based competition on result, Harvard Business Review Press, 2006. ISBN: 9781591397786.

[16] L.L. Weed, Medical records that guide and teach, N. Engl. J. Med. 278 (12) (1968) 652−657.

[17] J. Padikkaparambil, C. Ncube, K.K. Singh, A. Singh, Internet of things technologies for elderly health-care applications, Emergence of pharmaceutical industry growth with industrial IoT approach, Academic Press, 2020, pp. 217−243.

[18] M. Singh, S. Sachan, A. Singh, K.K. Singh, Internet of things in pharma industry: possibilities and challenges, Emergence of pharmaceutical industry growth with industrial IoT approach, Academic Press, 2020, pp. 195−216.

[19] M. Jangra, S.K. Dhull, K.K. Singh, Recent trends in arrhythmia beat detection: a review, in: B.M. Prasad, K.K. Singh, N. Ruhil, K. Singh, R. O'Kennedy (Eds.), Communication and computing systems, 1st ed., 2017, pp. 177−183.

[20] A. Mehrotra, S. Tripathi, K.K. Singh, P. Khandelwal, Blood vessel extraction for retinal images using morphological operator and KCN clustering, 2014 IEEE International advance computing conference (IACC), IEEE, 2014, pp. 1142−1146. February.

CHAPTER 8

Examining diabetic subjects on their correlation with TTH and CAD: a statistical approach on exploratory results

Subhra Rani Mondal and Subhankar Das

Researcher and Lecturer, Honors Programme, Duy Tan University, Da Nang, Vietnam

8.1 Introduction

Machine learning vision is the technology in which a PC can show and utilize at least one camcorder which is simple to advanced (ADC), and computerized sign for preparing (DSP) the analysis. The acquired information is constrained by the PC or robot. The location of a machine is used to find any future uncertain problems and to acknowledge their occurrence unpredictability to discourse acknowledgment. Each visual affect and goals framework has two significant highlights.

1. Sensitivity is the capacity of a gadget to identify shortcomings in powerless light or undetectable wavelengths.
2. Resolution is how much the gadget can contrast between articles. By and large, goals will in general break the point of the field of view. The affectability relies upon the straightforwardness. Every other factor stays consistent, lessening sharpness affectability and decreasing affectability goals.

The human eye is sensitive to electromagnetic radiation in the 390–770 nm range. Camcorders are delicate to a wide scope of wavelengths. A portion of the vision framework in this gadget works with infrared wavelengths, bright light, or X-rays. Sight requires a PC with a propelled processor. In detail, high-goals cameras require a lot of arbitrary access memory (RAM) and AI programming. Hardware scenes are utilized in different restorative and modern applications.

Here's a model:

1. Electronic segment investigation
2. Identify signature
3. Optical character acknowledgment
4. Handwriting acknowledgment

154 CHAPTER 8 Examining diabetic subjects on their correlation

5. Finding object
6. Pattern acknowledgment
7. Content investigation
8. Make cash
9. Medical picture investigation

The term gadget regularly alludes to mechanical PC programs; however, PC terms are frequently used to portray computerized PCs, information handling, and any sort of innovation for which some of them are recorded.

Segments of a machine vision system are unique, however, as a rule there are numerous components. These components are as follows:

1. Digital or simple camera to catch pictures
2. Instruments for digitizing pictures like camera interfaces
3. Processor

At the point when these three segments are associated with the gadget, it is known as a keen camera. The vehicle's visual framework can be gotten from savvy cameras outfitted with the accompanying additional items.

1. Input/yield equipment
2. Lens
3. Light sources, for example, LED lights and halogen bulbs
4. Image preparing program
5. Sensors to identify and access pictures
6. Parts classification

8.1.1 General application procedure

1. Product review
2. Visual stock control and the executives of standardized identification perusing, checking, interface store, and so forth.
3. The nourishment and drink industry utilizes car vision frameworks to screen quality.

In the medicinal field, machine vision frameworks are utilized in restorative imaging and examination strategies.

8.1.2 Medicinal imaging

Medicinal imaging alludes to those methods and procedures that are utilized to treat pictures of various pieces of the human body for analytic purposes and for the treatment of advanced well-being. The term medicinal imaging incorporates different radiography techniques.

1. X-rays
2. Fluorescence magnifying lens

3. Magnetic resonance imaging (MRI)
4. Medical ultrasound or ultrasound
5. Endoscopy
6. Elastography
7. Tactical imaging
8. Thermography

8.1.2.1 Therapeutic photography and connected imaging methods for positron emission tomography (PET)

Therapeutic imaging includes estimation and recording methods that create pictures, and in addition information that is regularly shown in outlines and maps. This incorporates electroencephalograms (EEGs), magnetoencephalography (MEGs), and electrocardiograms (EKGs).

8.1.2.2 How is medical imaging used in digital health?

Therapeutic imaging is significant for every restorative setting and all degrees of medication. Medicinal imaging empowers doctors to make progressively precise conclusions and suitable treatment choices. Without medicinal imaging, both analysis and treatment in advanced well-being cannot be exact at all levels.

Significant applications for medical imaging techniques include the use of:

1. Radiography images having fissures show obsessive changes in the lungs and explicit kinds of colon malignant growth.
2. Fluoroscopy makes sensible pictures of inward parts and human structures.
3. MRI sweep makes 2D pictures of the body and cerebrum.
4. Catching the two-dimensional picture of radiation discharged by an unblemished radioactive isotopic to identify scintigraphically dynamic natural regions, which might be related with an infection.
5. Positron emission tomography (PET): to distinguish and treat different ailments by utilizing explicit properties of isotopes and vigorous particles from radioactive material.
6. Therapeutic ultrasound: for imaging of incipient organisms, stomach organs, heart, chest, muscles, ligaments, conduits, and veins for clinical purposes.
7. Elastography indicates tractable properties of delicate tissue in the body.
8. The strategic picture of prostate, breast, vagina, pelvic floor bolsters the structure, and muscle development is accomplished by changing into an advanced picture.
9. Photo sound imaging: screens tumor angiogenesis, screens blood oxygen, analyzes cerebrum capacity, and analyzes skin melanoma.
10. Thermographic system for the conclusion of breat tumor utilizing applications, for example, thermography, contact thermography, dynamic angiography.
11. Tomography techniques: makes pictures of delicate organ structures (CT, PET outputs).

156 CHAPTER 8 Examining diabetic subjects on their correlation

12. Echocardiography: checks the precise structure of the heart, for example, the size of the room, the capacity of the heart, the heart valves, and pericardium.

8.1.2.3 Biomedical image and analysis

Biomedical imaging estimates the human body at different scales (magnifying instrument, microscope, etc.) [1]. They are estimated by different imaging techniques (for example, CT scanners, ultrasound gadgets, and so forth) and physical properties of the human body (remote thickness, X-ray turbidity, and so forth). These pictures are translated by experts (e.g., radiologists) to perform clinical work (e.g., analysis) and majorly affect doctor's choices. Biomedical pictures are normally 3D pictures, here and there with extra measurements (4D) and/or numerous channels (4–5D) (for example multidimensional MR pictures). As a clinical convention for stratification of picture procurement strategies (e.g., patients on the back not bowed), changes in biomedical pictures are altogether different from ordinary pictures (e.g., photos). In the investigation, we plan to recognize subtleties (e.g., some little zones showing irregular findings) [1].

8.1.3 Big data and Internet of Things

As the name suggests, Big data refers to a lot of information. However, that is not all. Notwithstanding the amount, IBM information researchers have perceived huge information to indicate assorted variety, speed, and exactness.

Big data is the consequence of an assortment of assets, including online networking, exchanges, authoritative substance, sensors, and cell phones. It deals with huge number of collected data. Like clockwork, a 72-hour video is transferred to YouTube, 216,000 posts are sent, and 204 million messages are sent. With certainty, the gathered information ought to be in great time always refreshed progressively. Huge scale information examination is of extraordinary incentive to organizations and people that utilization it.

Then again, the Internet of Things (IoT) changes "things" into items of great interest consistently. Transport compartments with sensors interconnect with the cooler, watch, indoor regulator, vehicle, web for information gathering, and transmission. This data, when joined with data from different sources and other enormous information above, can be huge data [2,3].

8.1.4 Artificial intelligence (AI) and machine learning (ML)

8.1.4.1 Artificial intelligence

The term AI includes two words: "Artificial" and "Intelligence." Artificial means human-made things and abnormalities, and intelligence means the ability to understand and think. There is a misconception that artificial intelligence is a system, but it is not a system. AI runs on the system. AI has many definitions, but the definitions are as follows. This is the intelligence that we want to add to give all the capabilities to a device [4,5].

8.1.4.2 Machine learning

Machine learning (ML) is a learning device that can be learn with its own explicit program. It is an AI application that allows the system to automatically learn and improve the experience. You can now combine the application inputs and outputs to create an application. A simple definition of machine learning is that it is an application of artificial intelligence which has the ability to automatically learn and improve from repeated exposure to the data programming. If class work learners measured by P are improved by experience, then T classes and performance measures P [6−8].

8.1.4.2.1 Utilization of machine intelligence in healthcare

Here are the best 14 uses of AI in human services:

1. Various mechanism for health insurance is something that is found in the steady acknowledgment of the healthcare sector. Google has as of late propelled an AI calculation to recognize malignant tumors in mammography, and Stanford scientists have utilized ML to figure out how to distinguish skin disease. ML right now has different obligations in well-being. ML enables medicinal services to investigate and examine a large number of various information foci, giving auspicious hazard appraisal, precise asset distribution, and different applications.
2. With the expanding utilization of machine learning in medicinal services, it will be conceivable to help a great many patients in future without having to bother for any sort of health check up data as it can study the huge set of existing data. Before long it will spread to ML-based projects that contain continuous patient information from numerous medicinal frameworks in different nations, which will expand the viability of new treatment alternatives. This was not accessible before [9−11].
3. Disease identification and diagnosis: one of the significant ML programs in human services is diagnosing illnesses that are generally hard to analyze. It can incorporate anything from malignancies that are hard to identify in the early stages to hereditary conditions.
4. Drug discovery and production: one of the clinical uses of AI is the early identification of medications. It likewise incorporates R&D advancements, for example, cutting-edge sequencing and restorative exactness to help discover options in contrast to the treatment of multifactorial ailments.
5. Currently, AI strategies incorporate unrivaled discovering that can recognize information designs without forecasts.
6. Imaging medicinal recognition: machine learning and profound learning are both in charge of a historic innovation called PC vision. The discoveries were affirmed by Microsoft for the Inner Eye activity, which works with

158 **CHAPTER 8** Examining diabetic subjects on their correlation

picture acknowledgment instruments for picture examination. As ML turns out to be progressively commonplace and its illustrative capacity expands, it is normal that more wellsprings of various therapeutic pictures that are a piece of this AI-based symptomatic procedure will be seen.

7. Personal care: personal treatment must be viable by pairwise singular well-being by prescient investigation, and also for further sickness research and assessment.

8. Machine learning change: tracks changes as preventive medication, and extension of mechanical preparation in medicinal services, counteractive action and malignancy identification with various patients' information.

9. Intelligent health records: keeping your health records state-of-the-art is a thorough procedure. Innovation is lessening the information transfer process, yet a large portion of the procedure still requires a great deal of energy. The primary job of ML in medicinal services is to improve the way toward improving time-, effort-, and cost-efficiency. OCR discovery advancements, for example, the Google Cloud Vision API and MATLAB's AI induces innovations are slowly becoming famous.

10. Clinical trials and research: machine learning has a few potential applications in the field of clinical trials and research. Clinical trials are expensive in terms of energy and cash, and as a rule they might take years. Utilizing ML-based prescient investigation to recognize candidates for clinical preliminaries, potential analysts can pool specialists from different information foci, for example, past visits to doctors, web-based life, and so on.

11. Crowd-sourced data collection: crowdsourcing is currently concerned with the restorative network, enabling specialists and professionals to get an abundance of data.

12. Crude health information significantly affects how medications are dealt with on the web. With the advent of IoT, the social insurance industry keeps on finding better approaches to utilize this information, to help with testing diagnostics, and to improve in general diagnostics and drugs.

13. Better radiation treatment: one of the most utilized uses of ML is in the radiology office. Therapeutic picture examination has numerous individual factors that can occur.

14. There are numerous sores, diseases, and so on that cannot be improved by utilizing complex conditions. ML-based calculations are utilized in different occasions, making it simple to discover and discover factors. The most widely recognized utilizations of ML in restorative picture examination are to characterize articles, e.g., sores into classifications, for example, ordinary or unusual injuries or nonlesions [12−14].

8.1.5 **Big data and IoT applications in healthcare**

The fundamental issue is that all patients, particularly in remote regions, cannot get restorative treatment or treatment in basic circumstances. It has had unsavory ramifications for individuals with regard to clinic and specialist administrations. Today, these issues are generally understood by utilizing new advances that utilize IoT gadgets to screen human services.

1. Health tracking: massive data and analytics beyond the internet (IoT) is an upheaval that can follow client insights and data. Despite weak feeling that can uphold the patient rest, beat, work out, step, etc, there are new remedial progressions that can control the patient's circulatory strain, beat oximeter, glucose screen, etc.
2. Cost reserve funds: large measures of information are the most ideal approach to spare the expense of emergency clinics. Prescient investigation tackles this issue by anticipating confirmation rates and allocating representatives. This decreases medical clinic ventures and really boosts the speculation potential. The protection business can keep patients from being hospitalized by verifying sterile trackers in wearables [15−17].
3. Higher risk support: With all emergency clinical records being digitalized, the data can be used to predict for similar cases in future. Return these patients to the medical clinic to recognize their incessant issues. Such understanding will give better mind to such experts and bits of knowledge to redress restorative activities to lessen their successive visits. This is an incredible method to keep up a rundown of in danger patients and give them expert consideration.
4. Human mistake counteractive action: Sometimes specialist doctors are also accounted for giving wrong medicines. As a rule, a lot of information can be utilized to examine client information and physician recommended medications to decrease such blunders. This has extraordinary potential for reducing mistakes and sparing lives. Such programming is a great device for specialists to be presented to numerous patients.
5. Healthcare advances: regardless of the present situation and the enormous scale, the huge scale of information can lead to logical and mechanical advances. For human services, computerized reasoning like Watson IBM can utilize information in only seconds to discover answers for different ailments. This advancement is progressing and will proceed with the measure of research gathered by Big Data. Not only would it have the option to give an exact plan, yet it in like manner offers a remarkable response for one's own personal problems. The accessibility of unsurprising examination of patients moving to a particular geographic area will think about patients in a similar region.

8.1.6 Diabetes and its types

Diabetes (DM) has been recognized as a disease of pancreas with faulty insulin regulation in body. Because of a lack of insulin lack, glucose levels increase and hyperglycemia is brought about via starch, fat, and protein. DM affects more than 200 million individuals around the world. The likelihood of diabetes is likely to

160 CHAPTER 8 Examining diabetic subjects on their correlation

increase in the coming years. DM can be have a wide range of characteristics. However, there are two primary kinds of facilities, type 1 diabetes (T1D) and type 2 diabetes (T-2D) [18,19]. The most widely recognized type of T2D diabetes covers 95% of all diabetes patients, primarily because of insulin opposition, lifestyle factors, physical action, dietary propensities, and innate coronary illness [20]. Due to the devastation of the T1D effective system, it is visible from planarians' pancreases beta cells intestlezes. T1D affects about 10% of diabetes patients around the world, 10% of which in the end create idiopathic diabetes. Gestational diabetes, endocrine issue, type 2 diabetes mellitus, neuritis, mitochondria, and pregnancy depend on the details of the different kinds of DM and the beginning of insulin purification.

8.1.6.1 After effects of diabetes

The long-term confusions of diabetes continuously increase. The more diabetes, then the less glycemic control, and thus the greater is the danger of more complexities. Potential entanglements are:

1. Cardiovascular disease: diabetes increases the danger of cardiovascular damage, for example, angina, heart attack, stroke, and blood vessel stenosis (atherosclerosis, e.g., coronary corridor disease).
2. Neuropathy (additional sugar content): this can fortify the dividers of little veins (vessels), particularly in the leg.
3. There is a danger of pain in limbs and internal organs. It normally begins with your finger and spreads gradually. Whenever left untreated, you can lose every one of your organs.
4. Damage to the gastrointestinal tract can cause queasiness, vomiting, diarrhea, or constipation. For men, this can prompt erectile dysfunction.
5. Kidney infection (nephropathy): there are a huge number of glomeruli in the kidney, which channel the blood. Diabetes can harm this sensitive filtration framework. Genuine damage can prompt kidney damage or irreversible kidney illness, which may require dialysis or kidney transplantation.
6. Eye harm (retinopathy): diabetes harms the retinal veins (diabetic retinopathy) and causes visual impairment. Sickness builds the danger of creating physical conditions, for example, cataracts and glaucoma. Since untreated buildups are decreased and inoculated, once in a while the treatment of genuine contamination happens. All things considered, these are irresistible breaks.
7. Skin condition: due to skin issues including bacterial and contagious contaminations, diabetes can happen.
8. Deafness: deafness in individuals with diabetes is increasingly normal.
9. Alzheimer's illness: type 2 diabetes can increase the danger of dementia, e.g., Alzheimer's disease.
10. Depression: symptoms of sorrow are normal among individuals with type 1 and 2 diabetes [21−23].

11. Pregnancy diabetes connections: most ladies experiencing gestational diabetes have a healthy child. Be that as it may, loss of glucose control can be an issue for you and your infant.
12. Pregnancy related diabetes can affect children.
13. Hypertrophy: excess glucose can pass through the placenta, leading to very enormous infants [24–26].
14. Hypoglycemia: children with gestational diabetes can create glucose (hypoglycemia) after birth because of high insulin generation. Intravenous glucose arrangements can reestablish glucose levels of the typical child.
15. Type 2 diabetes: the mother with gestational diabetes is in danger of stoutness and type 2 diabetes.
16. Gestational diabetes can cause the death of the child when born.

8.1.6.2 Diabetes and headache

Not every person diabetic will encounter a migraine. The individuals who have as of late been determined to have diabetes will in general have migraines since they are as yet attempting to control their glucose and utilize an eating routine. For individuals with diabetes, cerebral pains are as a rule because of changes in glucose levels. A migraine can show that your glucose is high, and your PCP calls it hyperglycemia. Rather, glucose levels might be extremely low, a specialist called it hypoglycemia. Changes in glucose levels are bound to cause cerebral pains for diabetics [27–29].

8.1.6.3 Obesity and overweight

Obesity and overweight are ailments that influence well-being. Specialists generally demonstrate that corpulent individuals are fat. Body mass index (BMI) is an equation used by doctors to evaluate whether an individual is appropriate for weight, age, or stature. A BMI in the range of 25–29.9 shows that the individual is overweight; in excess of 30 shows that an individual is obese. Eating excessively can increase weight and cause obesity. People who utilize an eating regimen of essentially entire grains, water, natural products and vegetables are in danger of being overweight. Regardless of having sound weight, they indulge in heavy eating practice. New nourishments and beans contain fiber, which can make others feel much improved and advance sound absorption [30–32].

8.1.7 Coronary artery disease (CAD)

If coronary supply routes are excessively restricted, cardiovascular heart disease (CHD) or coronary conduit infection will advance. Coronary supply routes are veins that convey oxygen and blood to the heart. CHD produces cholesterol in the blood vessel divider. These plaques cause blood vessel stenosis and reduce the bloodstream to the heart. Thrombosis can in some cases disturb the circulatory system and cause genuine medical issues. CHD can cause angina, which is a sort

162 CHAPTER 8 Examining diabetic subjects on their correlation

of chest agony related with coronary illness. Angina can cause the accompanying feelings across the chest:

1. Squeeze
2. Pressure
3. Weight
4. Tightening
5. Burn
6. Pain [33]

8.1.7.1 Treatment

There is no solution for CHD. Be that as it may, there are manners by which one can deal with the circumstance. Treatment incorporates changing your way of life, including stopping smoking, having a sound eating routine, and standard exercise. Be that as it may, there are individuals who need to get prescriptions and treatment.

1. Medicine: doctor discusses prescription. There are different medications for the treatment of cardiovascular sickness.
2. The following are the medications that individuals can use to diminish the hazard and effect of CHD:
   a. Beta-blockers: physicians may recommend beta-blockers to lower circulatory strain and heart rate, particularly for individuals who have had a past heart attack.
   b. Showers, nitroglycerin patches or pills: these build veins, expand pulses in the heart, and alleviate chest pain [33,34].

8.1.7.2 Insulin

Insulin is a hormone that enables cells to convey glucose from its sources. The pancreas is the primary driver of insulin levels which controls and regulates blood glucose [35]. At the point when this occurs, glucose remains in the blood and cells cannot assimilate it and cannot change over to glycogen if insulin is not regulated the process. This is the presentation of T1D, and individuals with this sort of diabetes need ordinary insulin to endure. In certain individuals, particularly overweight, corpulent or idle individuals, insulin isn't compelling at transmitting glucose and cannot upgrade its adequacy. Type 2 diabetes happens when insulin islets don't demonstrate insulin obstruction.

8.1.7.3 Hypertension

Hypertension (HTN/HBP) is a long-term ailment where pressure in the veins rises.

1. Complications: coronary conduit sickness, stroke.
2. Other names: blood vessel hypertension, hypertension.
3. Diagnostic technique: resting pulse.
4. Treatment: way of life change, compulsion.

8.1.7.4 Counteractive action

Type 1 diabetes cannot be stopped. Be that as it may, picking a sound way of life to forestall diabetes, type 2 diabetes, and gestational diabetes can anticipate it.

1. Eat healthy foods: choose low-calorie and low-calorie sustenance and increase fiber. Concentrate on natural products, vegetables, and entire grains.
2. More physical movement: the objective of normal physical action is 30 minutes in the day. Like cycling walking which will be progressive and worked up slowly on daily basis.
3. Intended of losing additional weight of 200 pounds (90.7 kg) and 7% of body weight for instance, if someone looses 14 pounds a month, you can diminish the danger of diabetes slowly but it has to be healthy.
4. However, don't attempt to get thinner during pregnancy. To keep your weight sound, center around ceaseless change in your eating regimen and exercise propensities. Move yourself by collecting the advantages of shedding extra pounds.
5. Sometimes medicine is an alternative. Oral diabetes medications, for example, metformin (e.g., glucophage, glutasezza) may decrease the danger of type 2 diabetes, yet you have to pick a sound way of life. Check your blood glucose levels at any rate once every year to guarantee that you don't create type 2 diabetes [36,37].

8.2 Review of literature

As indicated by researchers [38], sugar level can increase with one hour of mobile or tablet use in night.

Groups at the University of Strasbourg and the University of Amsterdam analyzed the impact of light blue (light discharged by the gadget) on glucose guidelines. Artifical lights having high lumionsity may create type 2 diabetes if exposed to it more at discos and pubs.

The University of Strasbourg and the University of Amsterdam are exploring for food eating and glucose resistance on the next day, by presenting male rodents to water-blue light for one hour around night. The mice utilized in this investigation were every day, implying that they were alert during the day and dozed around evening time.

Subsequently, the next day, the creatures were given the decision of a reasonable eating regimen (rat sustenance), water, pork, and water.

The Researchers have discovered that blue light is an hour even enough to eat more sugar during the evening. Since the retina is delicate to the ignored light of the gadget and sends data to the piece of the cerebrum that controls craving, specialists clarified that this relationship may exist.

They state the discoveries recommend that individuals who use telephones, tablets and PCs during the night are having weak defense mechanism against diabetes.

Anayanci Masís-Vargas, the lead of this study, said [39] "Restricting the time spent before the screen around evening time has been simply the most ideal approach to shield from the hurtful impacts of light blue." They worked on night projects and involve gadgets which emits orange and blue lights.

The discoveries were introduced to the Respiratory Society in the Netherlands (SSIB).

English investigations have demonstrated that the individuals with corpulence, hypertension and CAD are bound to smoke than the individuals who has no such cases in the United Kingdom. They gain weight and may get diabetes. Medications can help the government to handle obesity [40−42].

As indicated by analysts, about 33% of grown-ups in the United Kingdom are obese, Smoking is as yet the main source of disease in the nation, and the danger of sickness is greater in obesity.

Examination of disease in the United Kingdom has demonstrated that overweight and obesity are the fundamental drivers of four unique sorts of malignancy. In the United Kingdom, overweight and smoking causes around 1900 instances of colon disease. A similar example is identified with kidney malignant growth (multiple times in a year in the UK), ovary (460), and liver (180).

Mitchell [42], leader of the UK stated that malignant growth can be a result of smoking in youth.

At 300 calories a day [43], the degree of markers, for example, great cholesterol, circulatory strain, glucose and so forth will be incredibly improved. Grownups younger than 50 were on the eleventh July rundown of Lennesty's diabetes and endocrinology.

This investigation was led by the National Institute for Aging, Institute of Gastrointestinal Allergy (NIH Assistance U01AG022132, U01AG020478, U01AG020487, U01AG020480) and the NIH Comprehensive Clinical Research Center [43].

8.3 Research methodology

8.3.1 Trial setup

The subjects picked were experiencing diabetes as indicated by their Health Insurance Providers without uncovering any Personal Information (PI) or Sensitive Personal Information (SPI) by Law. To recognize each case particularly, test ids like S1, S2, and so forth were distributed to the subjects. We gathered this huge information and concentrated the individuals; we have considered their strain level and helped them to fix it. In this section, we did our best to investigate diabetes with CAD and different diseases. Test data from lab tests and medicinal reports have been gathered for the following attributes:

1. Gender
2. Age

3. Diabetes Type
4. Subject on Insulin
5. Subject having Obesity
6. Subject having CAD
7. Subject having HTN
8. Subject experiencing TTH (aka. cerebral pain)/Migraine.

Our zone of interest is on the TTH/Migraine parameter. The investigation centers around finding the job of diabetes in causing TTH and what are the unconventional probabilities/design which do/can lead the subject to TTH.

We are intrigued to check the example and conduct of the information pattern of subjects enduring CAD (Coronary Artery Disease). To discover the example and relationship on various parameters, we first, broke down the gathered example information.

The Tableau programming was utilized for the examination.

In the present analysis, we have gathered the example of 30 irregular Indian subjects and they were profoundly examined under different restorative parameters. These subjects were experiencing diabetes and their different side effects were recorded. It was checked whether there is plausibility of different sicknesses with diabetes. Patients with type 1 and II diabetes were explored along with the individuals who were and who were not taking insulin.

The outcomes demonstrate the connection of various different sicknesses like weight, hypertension, CAD, and so on with TTH and its immediate expectation of the occurrence of these issues in the future course of life. The age and sexual orientation parameters were likewise dealt with and confined examination was executed with respect to these (Fig. 8.1).

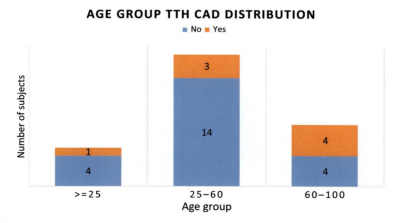

FIGURE 8.1

TTH-CAD distribution among subjects as per age groups.

8.4 Result analysis and discussion

The overall age group with the number of subjects with or without TTH and CAD is plotted in (Fig. 8.2). It can be observed that very few cases are observed for TTH for males below 60 but the number increased dramatically as age advanced above 60. This signifies that old-age men have the highest probability among all segments of having TTH and CAD. It also shows a direct correlation of TTH and CAD as per increasing age group (Fig. 8.2).

In the collected sample, in total 31.58% of subjects are reported as having TTH and CAD. This number becomes significant as it is saying that out every four diabetic patients one is suffering from TTH-CAD. Diabetic patients generally reported having TTH due to stress and mental demotivation caused due to problems like diabetes. On analyzing this ratio distribution within the gender, it has been found that males are more prone to TTH and consequently to CAD than females. Around 10.9% more males are reported as having TTH than females. On a detailed look into the data and subjected to discussion it has been found that the sampled males were working in an office environment while females are working less, so there is a possibility of work stress causing more frequent headaches (Fig. 8.3).

The overall age group with gender and number of subject with or without TTH and CAD is plotted in (Fig. 8.4). It can be observed that very few cases are observed for TTH for the males below 60 but the number increased dramatically as age advanced above 60. This signifies old-age men have the highest probability among all segments of having TTH. Also, for females increasing trends of TTH can be observed up to age of 60 but beyond that cases of TTH and CAD decrease. This is in line with the overall pattern of diabetic subject variation with age group (Fig. 8.4). In order to visualize this

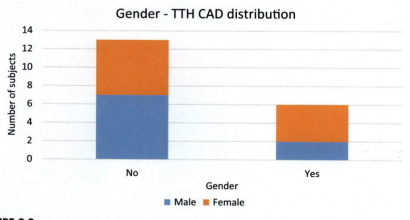

FIGURE 8.2

TTH-CAD distribution among subjects as per gender groups.

8.4 Result analysis and discussion

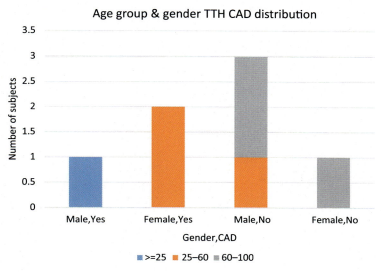

FIGURE 8.3
TTH-CAD distribution among subjects as per age groups in terms of presence of CAD in different genders.

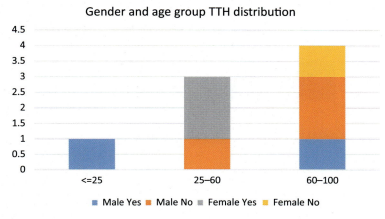

FIGURE 8.4
TTH-CAD distribution among subjects as per age groups in terms of presence of CAD.

pattern keenly, the upper graph is plotted only for TTH with CAD subjects and lower graph is obtained:

In conclusion, it can be said that TTH-CAD cases increase with age in males and do not follow the pattern of diabetes variation with age, while female

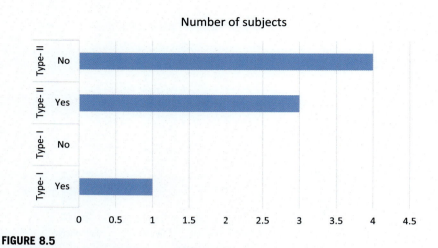

FIGURE 8.5

TTH-CAD distribution w.r.t. presence of diabetes type I and II among subjects.

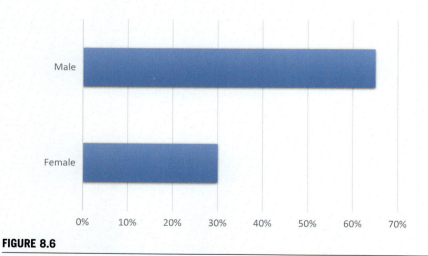

FIGURE 8.6

Correlation of diabetes with TTH and possibility of CAD, HTN, obesity etc. diseases with insulin Consumption.

TTH-CAD pattern variation is the same as diabetes, i.e., increasing trend up to age of 60 then decreasing (Fig. 8.5 and Fig. 8.6). In order to gain more insight and to know the role of diabetes type in causing TTH-CAD, a stacked bar relationship of diabetes type and TTH was plotted and it was observed that type 1 diabetes has almost no contribution to TTH and CAD. In other words, a subject suffering from type 1 diabetes reports much less TTH and CAD too. This could

8.4 Result analysis and discussion 169

Table 8.1 The summary data for the Diseases TTH & CAD as per type 1 and 2 diabetes.

Diabetes type	TTH	% of Total number of subjects	% of total number of subjects within each diabetic type	Number of subjects
Type 1	Yes	12.5	100	1
Type 1	No	0	0	0
Type 2	Yes	37.5	42.85	3
Type 2	No	50	57.15	4

be justified since type 1 subjects are of younger age groups so they are able to cope with tense situation or it might be the case that they do not have such a significant level of tension with coronary disease that can cause the TTH and CAD (Table 8.1). But the number increases significantly for type 2 diabetes patients. The summary data are as follow:

8.4.1 TTH cannot be

In the investigation into the different coexistence of illness causing TTH or not, it can been seen that the person who is on insulin and has CAD but no HTN is safe from TTH. In the samples the subjects who fall under these parameters are found to be in the 25−60 age group from both the genders. In such a combination male dominance can be easily seen. Two thirds of the population is male in this combination (Fig. 8.7). Such subjects are found to be sufferers of type 2 diabetes, so they are insulin consumers. Subjects do not have issues related with digestion systems as they were reported to have no obesity problems. So, it can be concluded that if patients are not having problems related with blood pressure then even if they have CAD then they cannot have TTH. This signifies that if subjects have control of their hypertensions and blood pressure then they can save themselves from TTH even with cardiac illness. When drilling down into the data it has been found that the ratio of females is less in such categories because the problem of depression is not reported in females, while male patients are equally distributed between depression and not depression. So we can say such category subjects are equally divided into three groups: female, male with depression, and male without depression. In the sample collected it has been seen that females havse negligible depression, which may be one of the major reasons why overall TTH is less in females as compare to males.

When analyzing the sample with TTH, an interesting result can be seen that the ratio of old-age males having TTH is maximum that can be easily seen in Age group gender distribution study but this graph shows they are insulin takers and they are having hyper blood pressure. So, it can be concluded clearly that if an old-age male has hypertension and is a sufferer of type 2 diabetes then he must have TTH. On a deeper look it shows that a 1:3 ratio is further found for

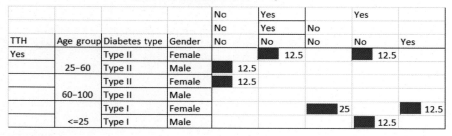

FIGURE 8.7

Summary chart of analytical data for correlation of diabetes with TTH and possibility of CAD, HTN, obesity etc. diseases with insulin Consumption.

such males having CAD and not having CAD and the same ratio is found for obesity as well (Fig. 8.7). Overall if considered then TTH subjects are generally having HTN, as 6 out of 8 Subjects which are reported of having TTH are having HTN as well i.e. 75% of the TTH subjects. These numbers clearly signify HTN has a relation with the TTH. If a subject has diabetes and HTN then one day or the other there are very high chances of TTH occurring as well and the probability is higher if the subject is male.

8.5 Originality in the presented work

The IoT use in healthcare helps us to add sensors to wearbale devices that can track pulse rate.

Mass data research in human services: Today, there are countless different assets, for example, interpersonal organizations, the Internet, and medicinal services frameworks. The utilization of a few innovation patterns, for example, the development of IoT gadgets help us to store data and use them for devising new machines.

All in all, medicinal services frameworks provide particular information that is hard to store, process, and translate. These data are organized, semiorganized, and unstructured [2,3], and it is difficult to oversee them with customary database. Accordingly, the utilization of a lot of information in health frameworks is profoundly required. Great information can be utilized to improve therapeutic quality with past medicinal information, and would now be able to improve restorative quality [15]. There are many AI strategies used to break down sensor information in health frameworks. These advances help to improve and are crafted by social insurance workers and screen medicinal services, to make the framework progressively precise. The point of our exploration is to avoid and treat diabetes and improve those living with diabetes, which can be achieved by improving the grouping technique.

Given the accessibility of the above data, more consideration is being paid to the social insurance industry and more research is being done to target information to improve medicinal services.

Health information is becoming progressively more complex with restorative staff, medicinal records, therapeutic records, clinical records, therapeutic imaging, digital physical frameworks, restorative web-related issues, hereditary information, and frameworks. The increase in clinical assistance and clinical choices is noteworthy.

New sorts of information from sources, e.g., interpersonal interaction administrations and genomics information, are utilized to manufacture customized individual consideration frameworks. Accordingly, various sorts of health information from various sources, fields, and advances, and their nature are subject to legitimate investigations. All scientific research needs to beat these impediments to mining information and produce significant experiences to save lives.

8.6 Future scope and limitations

We have found from our example examination that our example isn't powerless to sex. It tends to be utilized to dissect enormous example tests later on. We can without much of a stretch discover changes in TTH with changes in different infections in a subject separate from the sicknesses referenced, for example, weight, malignancy, and heart attack [32]. We utilized Tableau S/W for huge accumulations. This sorting of TTH and the two kinds of diabetes rely upon an assortment of components, for example, age, sexual orientation, type of diabetes, insulin issue, weight, subject to CAN, subject to TTH (aka. cerebral pain)/headache, and so on [28,36,44].

Since the research sample size is small, it can not give more accurate details. So more data and patient informations on type 1 diabetes can be more helpful. With the assistance of innovation, it is important to make a framework for the creation and examination of large-scale diabetes information and to anticipate the potential dangers dependent on it. Prescient examination is a procedure that joins different information mining strategies, AI calculations, and measurements that incorporate present and recorded informational indexes to pick up experiences and anticipate future dangers.

With side effects that might be due to low or high glucose, hypoglycemia or absence of information about drying out may be researched for further study. This investigation isn't planned for pregnant ladies, dialysis patients, or malignant growth patients.

8.7 Recommendations and considerations

The creator tried to incorporate age, sexual orientation, insulin elements, and associations with diabetes. Issues under investigation for diabetes under a title

172 **CHAPTER 8** Examining diabetic subjects on their correlation

given by medical coverage without revelation of individual data (PI) or delicate individual data (SPI) are liable to the law. At long last, TTH-CAD cases shift with age in men, and the absence of examples of diabetes changes with age, however changes in TTH-CAD designs in ladies are smiliar to those for diabetes, i.e., the pace of increase to 60 years old and consequent decay rates.

It very well may be seen that not very many cases are seen for TTH for males below 60, however the number expands significantly as age advances beyond 60. This indicates that old-age men have the most elevated likelihood of having TTH and CAD. It additionally demonstrates an immediate connection of TTH and CAD according to increasing age [25,36].

Additionally, for females increasing patterns of TTH can be seen up to age of 60, but past that instances of TTH and CAD diminish. This is in accordance with the general example of diabetic subject variety with age gathering.

It tends to be advocated that type 1 subjects are of more youthful age ranges so they can adapt to tense circumstances or it may be the situation they don't have such a noteworthy degree of pressure with coronary illness that can cause TTH and CAD. Be that as it may, the number increases essentially for type II diabetes-tolerant individuals.

In the example, subjects or patients are observed to be of the 25−60 age group from the two sexes who fall under these parameters. In such a blend male predominance can be effectively observed. Two thirds of the populace is male with this combination. When examining those having TTH, a fascinating outcome can be seen that the proportion of old-age males having TTH is greatest that can be effectively found in Age group sexual orientation dispersion study however this chart indicates they are insulin purchaser and they are having hyper pulse. In this way, it very well may be finished up obviously that if maturity male is having hypertension and sufferer of sort II diabetes then he should have TTH. On more profound look it demonstrates that there is 1:3 proportion is additionally found for such guys having CAD and not having CAD and same proportion is found for stoutness also.

8.8 Conclusion

In the present Analysis, we have gathered the conceivable example of 30 arbitrary Indian subjects and they were profoundly examined under different restorative parameters. These subjects were experiencing diabetes and their different side effects were recorded. It was checked whether there was a plausibility of different illnesses with diabetes. Patients with diabetes type 1 and 2 were examined along with the individuals who were and who were not taking Insulin. The outcomes demonstrates the relationship of various different sicknesses like weight, hypertension, CAD, and so forth with TTH and its immediate forecast of the occurrence of these issues in the future course of life. The age and sexual orientation parameters were additionally dealt. In the gathered example, 31.58% of subjects had TTH and CAD [33]. This number ends up huge as it is stating that out of every four diabetic patient one is experiencing TTH-CAD [27,45,46]. Diabetic patients are commonly revealed as having TTH because of stress and mental demotivation caused because of issues like diabetes [28,47].

On dissecting this proportion with respect to sexual orientation, it has been discovered that men are increasingly inclined to TTH and thus to CAD, compared to females. Around 10.9% more men are accounted for having TTH than females. The men tested were working in office conditions, while the females were working less so there is the possibility of work pressure which causing increasing cerebral pain.

Taking everything into account, it very well may be said that TTH-CAD cases rise with age.

In general TTH subjects are by and large having HTN. Six out of eight subjects who have TTH have HTN too, i.e., 75% of the TTH subjects. These numbers plainly imply that HTN has a connection with the TTH. If a subject has diabetes and HTN, there are high odds of TTH happening too and the likelihood increases if the subject is male.

This chapter has discussed new machine learning calculations, enhancement calculations, and health applications. There are significant issues, for example, security, prestudies, genuine ventures, and connections between information investigation and human services staff. These issues are fundamental for medical services improvement, otherwise it is hard to discharge AI calculations and upgrade real activity.

Accordingly, people are relied upon to advance human services designs in the coming decades. To secure future patient well-being and access to future health administrations, it is important to guarantee that health administrations are financially maintainable and practical.

Proof was found for research endeavors planned for creating man-made reasoning instruments to forestall and avoid entanglements related with diabetes. Our discoveries recommend that man-made consciousness innovation is appropriate for use in clinical practice. Accordingly, these techniques are incredible assets for improving the personal satisfaction of patients.

References

[1] M.S. Abdelhmid Salih, A. Abraham, Novel ensemble decision support and health care monitoring system, Int. J. Comput. Inf. Syst. Ind. Manag. Appl. (2014) 41−52.

[2] Arora, N., Rastogi, R., Chaturvedi, D.K., Satya, S., Gupta, M., Yadav, V., et al. (2019b), Book chapter titled as 'chronic TTH Analysis by EMG & GSR biofeedback on various modes and various medical symptoms using IoT', Paperback ISBN: 9780128181461, Chapter 5, Page No. 87-149, Advances in ubiquitous sensing

applications for healthcare, book-big data analytics for intelligent healthcare management. https://doi.org/10.1016/B978-0-12-818146-1.00005-2.

[3] Brazier, Y. (2018). *What is obesity and what causes it?*, Medical News Today, Retrieved from, https://www.medicalnewstoday.com/articles/323551.php 2nd November 2018.

[4] Chaturvedi, D.K., Rastogi, R., Satya, S., Arora, N., Saini, H., Verma, H., et al. (2018a) *Statistical analysis of EMG and GSR therapy on visual mode and SF-36 scores for chronic TTH*, in the proceedings of UPCON-2018 on 2−4 November 2018 MMMUT Gorakhpur, UP.

[5] Available at www.techrepublic.com/article/understanding-the-differences-between-ai-machine-learning-and-deep-learning.

[6] Arora, N., Trivedi, P., Chauhan, S., Rastogi, R., Chaturvedi, D.K. (2017a) *Framework for use of machine intelligence on clinical psychology to study the effects of spiritual tools on human behavior and psychic challenges*, Proceedings of NSC-2017 (National system conference), DEI, Agra, Dec. 1-3, 2017.

[7] Chaturvedi, D.K., Rastogi, R., Arora, N., Trivedi, P., Mishra, V. (2017b) *Swarm intelligent optimized method of development of noble life in the perspective of Indian scientific philosophy and psychology*, Proceedings of NSC-2017 (National System Conference), DEI Agra, Dec. 1-3, 2017.

[8] F.G. Cunningham, Diabetes mellitus, 24th ed., Williams obstetrics, 2014, The McGraw-Hill Companies, New York, NY, 2018. Available from: https://accessmedicine.mhmedical.com/. Accessed March 6, 2018.

[9] Felman, A. (2018). *An overview of insulin*, Medical News Today, Retrieved from https://www.medicalnewstoday.com/articles/323760.php, November 2018.

[10] S.G. Gabbe, Diabetes mellitus complicating normal pregnancy, 7th ed., Obstetrics: normal and problem pregnancy, 2018, Saunders Elsevier, Philadelphia, PA., 2018. Available from: https://www.clinicalkey.com. Accessed Jan. 10, 2018.

[11] M. Gulati, R. Rastogi, D.K. Chaturvedi, P. Sharma, V. Yadav, S. Chauhan, et al., Statistical resultant analysis of psychosomatic survey on various human personality indicators: statistical survey to map stress and mental health, Chapter 22 of handbook of research on learning in the age of transhumanism, IGI Global, Hershey, PA, 2019, pp. 363−383. Available from: http://doi.org/10.4018/978-1-5225-8431-5.

[12] Gupta, M., Rastogi, R., Chaturvedi, D.K., Satya, S., Arora, Verma, H., et al. (2019) *Comparative study of trends observed during different medications by subjects under EMG & GSR biofeedback*, ICSMSIC-2019, ABESEC, Ghaziabad. 8−9 March 2019. IJITEE, Vol. 8, 6S, pp. 748−756. https://www.ijitee.org/download/volume-8-issue-6S.

[13] Tsai, H.C.Cohly, H., Chaturvedi, D.K. (2013). *Towards the consciousness of the mind*, *towards a science of consciousness*, Dayalbagh Conference Proceeding, Agra, India.

[14] M.-H. Kuo, T. Sahama, A.W. Kushniruk, E.M. Borycki, D.K. Grunwell, *Health big data analytics: current perspectives, challenges and potential solutions*, Int. J. Big Data Intell. 1 (1-2) (2014) 114−126.

[15] Chaturvedi D.K., Lajwanti, T. Chu, H., Kohli H.P. (2012c). *Energy distribution profile of human influences the level of consciousness*, towards a science of consciousness, Arizona Conference Proceeding, Tucson, Arizona.

[16] Duke University Medical Center. (2019, July 11). *Even in svelte adults, cutting about 300 calories daily protects the heart*, researchers seeking a signal in metabolism or a 'magic molecule' to explain this. ScienceDaily. Retrieved July 19, 2019 from www.sciencedaily.com/releases/2019/07/190711183758.htm.

References   175

[17] R. Martin, S. Ira Ktena, N. Pawlowski, An introduction to biomedical image analysis with tensorflow and DLT, Imperial College London, 2018. Jul. 3.

[18] D.L. Kasper, *Diabetes mellitus: diagnosis, classification and pathophysiology*, 19th ed., Harrison's principles of internal medicine, 2015, McGraw-Hill Education, New York, NY, 2015. Available from: https://accessmedicine.mhmedical.com/. Accessed April 16, 2018.

[19] Masís-Vargas, A. (2019a). '*Society for the study of ingestive behavior.* Blue light at night increases the consumption of sweets in rats', Science Daily (2019, July 9). Retrieved July 19, 2019 from www.sciencedaily.com/releases/2019/07/190709091120.htm.

[20] Minn, R., Morrow, E.S., Allscripts, EPSi. (2018). '*Natural medicines in the clinical management of diabetes*', Natural Medicines. https://naturalmedicines.therapeuticresearch.com. Accessed March 6, 2018. Mayo Clinic, Accessed Jan. 17, 2018.

[21] R.A. Marrie, R. Patel, C.R. Figley, J. Kornelsen, J.M. Bolton, L. Graff, et al., Diabetes and anxiety adversely affect cognition in multiple sclerosis, Multiple Scler. Relat. Disord. Satell. 27 (2019) 164−170. Jan.

[22] Nordqvist, C. (2019). '*What to know about coronary heart disease*', Medical News Today, https://www.medicalnewstoday.com/articles/184130.php, 5th July 2019.

[23] T. Panch, P. Szolovits, R. Atun, *Artificial intelligence, machine learning and health systems*, J. Glob. Health 8 (2) (2018) 020303. Available from: https://doi.org/10.7189/jogh.08.020303. Published online 2018 Oct 21.

[24] Plant,L., Noriega,B., Sonti, A., Constant, N., Mankodiya, K. (2016). '*Smart E-textile gloves for quantified measurements in movement disorders*', In Proceedings of the IEEE MIT undergraduate research technology conference (URTC), pp. 1−4, 2016.

[25] Rachel Nall, M.S.N. (2018, November 8). '*An overview of diabetes types and treatments*', Medical News Today, Retrieved from https://www.medicalnewstoday.com/articles/323627.php.

[26] Rahmani, A.-M., Thanigaivelan, N.K., Gia, T.N., Granados, J., Negas, B., Liljeberg, P., et al. (2018d) '*Analytical comparison of efficacy for electromyography and galvanic skin resistance biofeedback on audio-visual mode for chronic TTH on various attributes*', in the proceedings of the ICCIDA-2018 on 27 and 28th October 2018, CCIS Series, Springer at Gandhi Institute for Technology, Khordha, Bhubaneswar, Odisha, India.

[27] Cancer Research U.K. (2019, July 2). '*Obese people outnumber smokers two to one*', ScienceDaily. Retrieved July 19, 2019 from www.sciencedaily.com/releases/2019/07/190702211335.htm.

[28] Masís-Vargas, A. (2019b). *One hour of device screen time at night could lead to increased sugar consumption*, at the society for the study of ingestive behavior (SSIB) in the Netherlands (2019, Wed, 10 July).

[29] Satya, S., Rastogi, R., Chaturvedi, D.K., Arora, N., Singh, P., Vyas, P. (2018a) *Statistical analysis for effect of positive thinking on stress management and creative problem solving for adolescents*, Proceedings of the 12th INDIA-Com; 2018 ISSN 0973-7529 and ISBN 978-93-80544-14-4, pp. 245-251.

[30] Rastogi, R., Chaturvedi, D.K., Satya, S., Arora, N., Singhal, P., Gulati, M. (2018a) Statistical resultant analysis of spiritual & psychosomatic stress survey on various human personality indicators, in the international conference proceedings of ICCI 2018. doi: 10.1007/978-981-13-8222-2_25.

[31] Rastogi, R., Chaturvedi, D.K., Satya, S., Arora, N., Yadav, V., Chauhan, S., et al. (2018b). *SF-36 scores analysis for EMG and GSR therapy on audio, visual and audio visual modes for chronic TTH*, in the proceedings of the ICCIDA-2018 on 27 and 28th October 2018 CCIS series, Springer at Gandhi Institute for Technology, Khordha, Bhubaneswar, Odisha, India.

[32] H. Saini, R. Rastogi, D.K. Chaturvedi, S. Satya, N. Arora, M. Gupta, et al., An optimized biofeedback EMG and GSR biofeedback therapy for chronic TTH on SF-36 scores of different MMBD modes on various medical symptoms, Chapter 8 of hybrid machine intelligence for medical image analysis, 841, Springer Nature Singapore Pte Ltd., 2019ISBN:978-981-13-8929-0. Available from: https://doi.org/10.1007/978-981-13-8930-6_8.

[33] Satya, S., Arora, N., Trivedi, P., Singh, A., Sharma, A., Singh, A., et al. (2019b). *Intelligent analysis for personality detection on various indicators by clinical reliable psychological TTH and stress surveys*, in the proceedings of CIPR 2019 at indian institute of engineering science and technology, Shibpur on 19th−20th January 2019, Springer-AISC Series.

[34] Saini, H., Rastogi, R., Chaturvedi, D.K., Satya, S., Arora, N., Verma, H., et al. (2018). *Comparative efficacy analysis of electromyography and galvanic skin resistance biofeedback on audio mode for chronic TTH on various indicators*, in the proceedings of ICCIIoT-2018, 14−15 December, 2018 at NIT Agartala, Tripura, ELSEVIER-SSRN Digital Library (ISSN 1556-5068). https://ssrn.com/abstract = 3354371.

[35] A. McAfee, E. Brynjolfsson, T.H. Davenport, D.J. Patil, D. Barton, Big data: the management revolution, Harv. Bus. Rev. 90 (10) (2012) 60−68.

[36] Rastogi, R., Chaturvedi, D.K., Satya, S., Arora, N., Chauhan, S. (2018c). *An optimized biofeedback therapy for chronic TTH between electromyography and galvanic skin resistance biofeedback on audio, visual and audio visual modes on various medical symptoms*, in the national Conference on 3rd MDSC PDR-2018 at DEI, Agra On 06-07 September, 2018.

[37] Rastogi, R., Chaturvedi, D.K., Satya, S., Arora, N., Sirohi, H., Singh, M., et al. (2018d). *Which one is best: electromyography biofeedback efficacy analysis on audio, visual and audio-visual modes for chronic TTH on different characteristics*, in the proceedings of ICCIIoT- 2018, 14−15 December 2018 at NIT Agartala, Tripura, ELSEVIER- SSRN Digital Library (ISSN 1556-5068). https://ssrn.com/abstract = 3354375.

[38] Sharma, S., Rastogi, R., Chaturvedi, D.K., Bansal, A., Agrawal, A. (2018). *Audio visual EMG & GSR biofeedback analysis for effect of spiritual techniques on human behavior and psychic challenges*, Proceedings of the 12th INDIACom; 2018, ISSN 0973-7529 and ISBN 978-93-80544-14-4, pp. 252−258.

[39] Singhal, P., Rastogi, R., Chaturvedi, D.K., Satya, S., Arora, N., Gupta, M., et al. (2019b) *Statistical analysis of exponential and polynomial models of EMG & GSR biofeedback for correlation between subjects medications movement & medication scores*, ICSMSIC-2019, ABESEC, Ghaziabad, 8−9 March 2019, IJITEE, vol. 8, 6S, pp. 625−635. https://www.ijitee.org/download/volume-8-issue-6S.

[40] A. Singh, R. Rastogi, D.K. Chaturvedi, S. Satya, N. Arora, A. Sharma, et al., Intelligent personality analysis on indicators in IoT-MMBD enabled environment, Chapter 7 of multimedia big data computing for IoT applications: concepts, paradigms, and solutions, Springer Nature Singapore, 2019. Available from: https://doi.org/10.1007/978-981-13-8759-3_7.

[41] Tenhunen, H. (2015). *Smart e-health gateway: bringing intelligence to Internet-of-Things-based ubiquitous healthcare systems*, In proceedings of the annual IEEE consumer communications and networking conference. NV, USA: IEEE, January 2015.

[42] Vyas, P., Rastogi, R., Chaturvedi, D.K., Arora, N., Trivedi, P., Singh, P. (2018). *Study on efficacy of electromyography and electroencephalography biofeedback with mindful meditation on mental health of youths*, Proceedings of the 12th INDIA-Com; 2018 ISSN 0973-7529 and ISBN 978-93-80544-14-4, pp. 84−89.

[43] V. Yadav, R. Rastogi, D.K. Chaturvedi, S. Satya, N. Arora, V. Yadav, et al., Statistical analysis of EMG & GSR biofeedback efficacy on different modes for chronic TTH on various indicators, Int. J. Adv. Intell. Paradig. 13 (1) (2018) 251−275. Available from: https://doi.org/10.1504/IJAIP.2019.10021825.

[44] American Diabetes Association (2018).*Dietary supplements*, http://www.diabetes.org/living-with-diabetes/treatment-and-care/medication/other-treatments/herbs-supplements-and-alternative-medicines/talking-to-your-health-care-provider.html. Accessed April 16, 2018.

[45] American Diabetes Association, Standards of medical care in diabetes, Diabetes Care. 34 (Suppl 1) (2011) S11−S61. Available from: https://doi.org/10.2337/dc11-S011.

[46] Bansal, I., Rastogi, R., Chaturvedi, D.K., Satya, S., Arora, N., Yadav, V. (2018) *Intelligent analysis for detection of complex human personality by clinical reliable psychological surveys on various indicators*, in the national Conference on 3rd MDNCPDR-2018 at DEI, Agra On 06-07, September, 2018.

[47] T.B. Murdoch, A.S. Detsky, The inevitable application of big data to health care, J. Am. Med. Assoc. (13)(2013) 1351−1352.

CHAPTER

Cancer prediction and diagnosis hinged on HCML in IOMT environment

9

G. S. Pradeep Ghantasala[1], Nalli Vinaya Kumari[2] and Rizwan Patan[3]

[1]*Department of Computer Science and Engineering, Chitkara University Institute of Engineering & Technology, Chandigarh, India*
[2]*Department of Computer Science and Engineering, Malla Reddy Institute of Technology and Science, Hyderabad, India*
[3]*Department of Computer Science and Engineering, Velagapudi Ramakrishna Siddhartha Engineering College, Vijayawada, India*

9.1 Introduction to machine learning (ML)

One of artificial intelligence's features is machine learning, which provides a framework to obtain information on mechanical failure. The focus of machine learning is on designing computer programs that input and utilize information for learning.

The creation of learning starts with clarifying or giving information, for example, direct experience or practice, to look at data trends and to take decisions in the perspective based on models. The most important reason is to permit the computers to be trained and to change behavior through design without human involvement and help (Fig. 9.1).

9.1.1 Some machine learning methods

Supervised machine learning algorithms are able to be applied to new data using tag examples to predict future events. The learning algorithm begins with an analysis of an established training dataset and provides a lesson in predicting the output quality. After sufficient training the framework will set objectives for a novel input. The learning algorithm can also match up to its output with either the expected output or detect miscalculations to adjust its form (Fig. 9.2).

9.1.2 Machine learning

Unsupervised master learning algorithms were secondary where there were no secret or marked sequences used for editing — unsupervised research studies on how systems could draw work from unwritten information to explain a hidden structure. The machine is not capable of finding the right output, but is searching the records and can copy data sets of unlabeled information to convey ritual structures.

Machine Learning and the Internet of Medical Things in Healthcare. DOI: https://doi.org/10.1016/B978-0-12-821229-5.00004-5
Copyright © 2021 Elsevier Inc. All rights reserved.

179

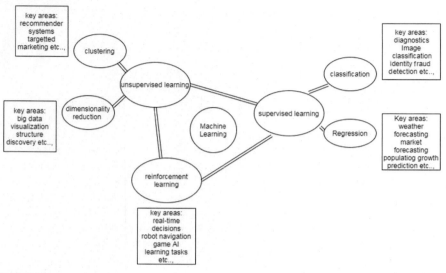

FIGURE 9.1

Machine learning.

Machine learning							
Supervised learning		Unsupervised learning		Semi-supervised learning		Reinforcement learning	
Continuous target variable	Categorical target variable	target variable not available		Categorical target variable		Categorical target variable	target variable not available
Regression	Classification	Clustering	Association	Classification	Clustering	Classification	Control
e.g., housing price prediction	e.g., medical imaging	e.g., customer segmentation	e.g., market basket analysis	e.g., text classification	e.g., GPS data	e.g., optimized marketing	e.g., driverless cars

FIGURE 9.2

Machine learning types.

Semisupervised machine learning algorithms require supervised and unlabeled learning wherever they are involved—usually a miniature quantity of marked data and a great number of unlabeled data. The devices using this method can achieve significant learning precision. Semimonitored education is usually chosen if the marked information obtained involves a high and related input for learning/learning. If not, it usually requires no more resources to acquire unlabeled data.

Algorithms of reinforcement teaching are schemes of information which communicate with their environment by procedures that detect errors or pillage. The most important aspects for enhancing training are trial and error quests and overdue repayment. This approach enables devices and computer agents to decide continuously on the best ways to make full use of their routine within an exact context. For the agent to be trained it is important to have clear reward input; whatever behavior is best, the stronger signal is recognized.

Machine learning allows the study of huge data volumes. Although planning to identify potential opportunities or risky risks usually produces more rapid and accurate results, it may need supplementary time to properly develop it. Combining machine learning with AI and cognitive technology can yet create additional performance in a broad sequence.

9.2 Introduction to IOT

The Internet of Things (IoT) is a scheme of consistent computational, automated, and digital machines, substances, animals, and people with distinctive IDs and the facility to transmit data through a network devoid of the use of an interface between people.

A human being with a cardiac monitor implant, and a mammal with a biochip transponder, an automobile machine which has built-in sensors to alert the driver to low tire stress, or some other natural or human-made entity which is able to provide an IP address to move data over a network can be an internet problem (Fig. 9.3).

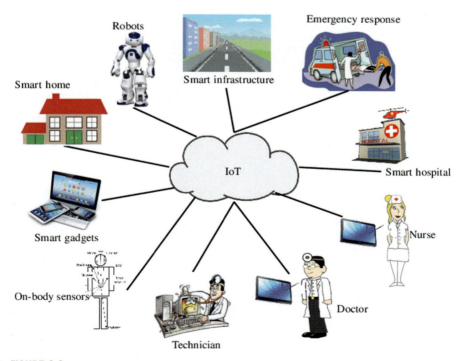

FIGURE 9.3

IoT software.

The use of IoT software in the current healthcare system is a benefit for doctors and clinicians because it operates in different areas of health care: real-time surveillance, patient information management, and health care. One of the main developments in IoT is the healthcare system.

Nonetheless, the introduction of this new technology in healthcare systems, without considering protection, leaves patients vulnerable to confidentiality. First of all, this paper highlights the key safety requirements of the BSN-based modern healthcare system. Then we introduce a robust IoT-based BSN medical care system, known as BSN-Care, which can meet these requirements efficiently.

9.3 Application of IOT in healthcare

Patient connections with physicians were restricted before the IoT in visits and experiences with television and media. Doctors and clinics were unable to continuously track or establish guidelines for patients' safety. The IoT strategy allowed remote monitoring within the health sector, opened up the potential to protect and improve patients, and empowered physicians to provide superlative treatment. It has also improved the appointment and acceptance of patients as contact with doctors has become more convenient and resourceful. Furthermore, mobile health monitoring helps to reduce the hospital time (Fig. 9.4).

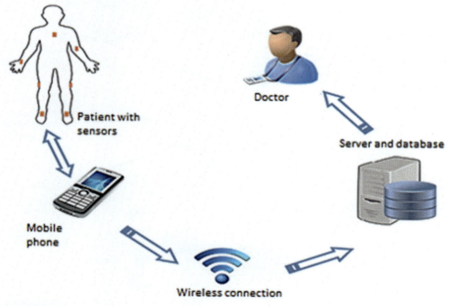

FIGURE 9.4

Patient monitoring system.

9.3 Application of IOT in healthcare 183

IoT for patients can be in the form of mobile, appropriate tape, and other instruments wirelessly compatible with the use of blood pressure and cardiac monitoring, glycometer, etc. Such machines can be used to calculate calories regularly, to conduct tests, appointments, changes in blood pressure, and much more.

By making a generic record of health conditions, IoT has contaminated people's lives, in particular elderly patients. This hurts people and their families most notably. Warning machinery sends a signal to connections about any disturbance or changes in a person's practice.

IoT for doctors can be used for patient's increasingly productive utilization of wearable and other home observing apparatuses implanted in IoT. You may screen patients' consistence with treatment plans or see any prompt human services needed. The IoT causes medicinal experts to be increasingly proactively cooperating with patients. The accumulation of information from IoT gadgets will help specialists to distinguish the best treatment for patients and to accomplish potential outcomes.

The IoT for Clinics—There are numerous different ways in + which IoT applications are helpful in medical clinics, aside from observing the well-being of patients. IoT edits the sensor tag to track medical devices "real-time situations, for example, wheelchairs, defibrillators, nebulizers, oxygen siphons and other viewing gears."

Improving contamination is a major problem for emergency clinical patients. Clean IoT-enabled viewing gadgets to stop contaminated patients. The IoT gear similarly promotes a resource on board, for example, a medical guide and environmental view, for example, a temperature cooling review and temperature control.

IoT for medical coverage organizations wellbeing back up plans with shrewd IoT applications likewise have impetuses. Insurance agencies may control information gathered from their guaranteeing and guarantee tasks through wellbeing checking devices. Such information will enable you to recognize trick claims and prospects for marks. IoT apparatuses permit the endorsing, evaluating, direct of cases and hazard appraisal forms simpler among back up plans and customers. Inside every single working procedure, customers are satisfactorily clear behind each choice made and framework results for essential idea, in spite of IoT-captained information driven choices.

Insurers can provide their customers with incentives to use and distribute IoT devices for health data. We are willing to compensate consumers for using IoT tools to monitor their daily behaviors and comply with their treatment plans and health prevention measures. It helps insurers to significantly reduce claims. In addition, IoT devices allow insurance companies to verify claims using the data collected by such devices.

9.3.1 Redefining healthcare

The explosion of IoT objects in the social insurance sector offers great opportunities. Also, the vast amount of information generated by these related gadgets will probably improve human services.

IoT has a four-stage structure on which the framework is based. Every one of the four stages are connected such that they record or process data over a period and return the incentive to the following stage. The qualities incorporated into the procedure offer instinctive and dynamic possibilities for the business.

Stage 1: The initial step includes the utilization of good gadgets including sensors, actuators, screens, identifier frameworks, camera frameworks, and so on. This information is gathered by these gadgets.

Stage 2: every now and again, sensor-perceived information and different techniques must be totaled in a simple structure and changed into computerized data for further information administering.

Stage 3: Information is digitalized, collected, and prehandled, indistinguishable and moved to the cloud or server farm.

Stage 4: Information at the compulsory level is overseen and broke down. Higher examination, utilized for these information, gives useful bits of knowledge into effective basic leadership.

IoT reclassifies human services through better social insurance, higher clinical outcomes and lower patient rates, better methodology and work processes, improved correspondence and patient associations (Fig. 9.5).

The major health benefits of IoT include:

- **Price decline:** IoT permits patient checking progressively, lessening downtime visits to doctors, emergency clinic stays, and reconfirmations essentially.
- **Improved treatment:** it empowers doctors to settle on educated and proof-based choices and guarantees full straightforwardness.
- **Fast analytic of sickness:** constant patient reconnaissance and information from continuous findings.

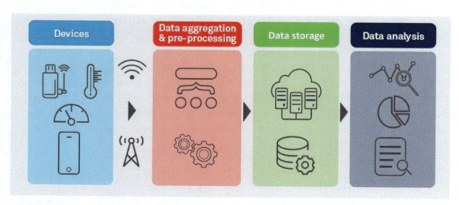

FIGURE 9.5

IoT solutions' four phases.

- **The association of medications and supplies:** the most significant issue for the social insurance industry is the dissemination of meds and medicinal gear.
- **Error decrease:** information produced by IoT gadgets permits wasteful choices as well as guarantees smooth social insurance with lower mistakes, misuse, and framework costs. Such frameworks can be successfully overseen and utilized by comparative gadgets with decreased expenses.
- **IoT isn't without hindrances for social insurance:** IoT-empowered associated gadgets gather huge information volumes, tally delicate information, and have information security concerns.
- **It is necessary to carry out adequate security measures:** through real-time medical surveillance and access to patient health data, IoT introduces a new field of patient care. Such information is good for healthcare workers to improve the health and understanding of the patient when taking advantage of opportunities and enhancing healthcare.
- It would make a difference in the increasingly connected world if we were prepared to leverage this electronic energy.

9.4 Machine learning use in health care

Machine learning is intended to make the machines more powerful, efficient, and successful than ever before. Nevertheless, the machine learning device is the mind and knowledge of the physician in a medical system.

A person needs a human presence and care all the time. This cannot be substituted for either machine learning or other hardware. The auditor can be better supplied by an automated machine. Below are listed the top 10 machine learning applications in health care.

9.4.1 Diagnose heart disease

The heart is one of the body's primary organs. Often, we have a variety of heart diseases like CAD, CHD, etc. We suffer from a wide range of heart diseases. Most scientists use algorithms for learning machines to diagnose cardiovascular diseases. It's a huge problem for science worldwide. One of the most excellent recompenses of machine learning is a mechanical heart disease diagnostics program.

Researches use multiple supervised machine learning algorithms as learning algorithms to detect cardiovascular diseases, such as the SVM or the Naive Bayes.

The UCI Heart Disease Dataset can be old or both training or testing data. For data analysis, the WEKA data mining software can be used in second-hand form. Additionally, you may be able to extend the process of diagnosis of heart disease through an Artificial Network (ANN) if you want (Fig. 9.6).

186 CHAPTER 9 Cancer prediction and diagnosis hinged on HCML

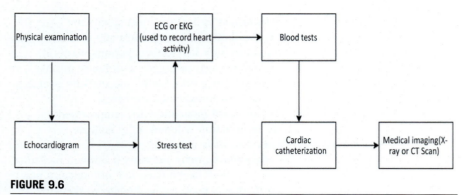

FIGURE 9.6

Diagnose for heart disease.

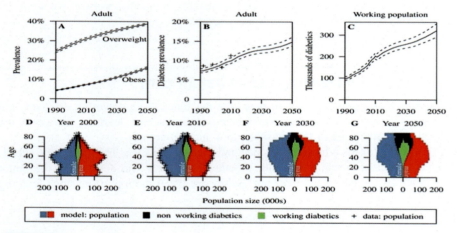

FIGURE 9.7

Diabetes prediction.

9.4.2 Diabetes prediction

One of the normal illnesses is diabetes. The condition is also one of the leading causes of other serious illness and death. The disease can affect different areas of the body such as the kidney, heart, and nerves. The goal is to diagnose diabetes early and to save patients by using a computer (Fig. 9.7).

Random forestry, KNN, Decision Tree, and Naive Bayes can be used for extending the diabetes forecasting method as an organizational algorithm. Naive Bayes gives consistency to the other algorithms. Since it is excellent and takes less time to add, the data set for diabetes can be downloaded. There are 768 data points, each of which has nine skin tones.

9.4.3 Liver disease prediction

The liver is our body's second most important internal organ. During metabolism it plays an important role. Most conditions such as cirrhosis, hepatitis chronic, liver cancer, and so on can be targeted.

Machine learning and data mining techniques for the prediction of liver disease were used radically. The prediction of diseases using large medical data is an unbelievably difficult task. Scientists were frustrated by problems with machine learning, such as cataloging, clustering, and much more (Fig. 9.8).

Indian Patient Liver Dataset (ILPD) may be used for the estimation of liver disease. There are 10 variables in this dataset. The Support Vector Machine (SVM) can be used as a classification system. The liver disease prediction model can be used by MATLAB.

9.4.4 Surgery on robots

Robotic surgery is one of the standard apps in health care for machine learning. In this context, there are four subcategories: mechanical suturation, surgical valuation, robotic surgical material design, and surgical modeling of workflow.

Cleft is the means by which an open wound is sewed. Suturing automation can condense the duration of surgery and fatigue of a surgeon. The Raven Surgical Robot is a phenomenon. Investigators seek to test surgery in automated, minimally persistent surgery for a machine learning development (Fig. 9.9).

Robots cannot operate in the case of neurosurgery. It takes time to manually work and can't automatically provide input. It can accelerate the process by using machine learning.

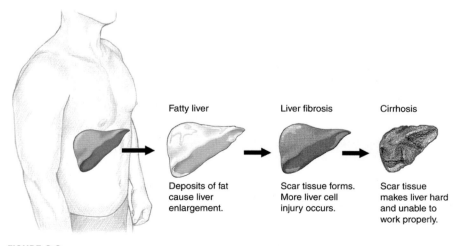

FIGURE 9.8

Liver disease prediction.

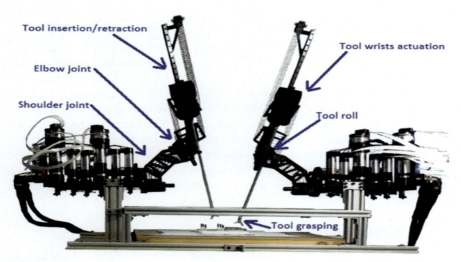

FIGURE 9.9

Raven surgical robot.

9.4.5 Detection and prediction of cancer

Machine learning methods are widely used for comprehensive tumor identification and classification in the identification of cancer. Deep analysis also plays an important role. When profound information can be accessed conveniently, there are data sources. A study showed that mistakes of diagnosing breast cancer are minimized by deep learning (Fig. 9.10).

Machine learning demonstrated its ability to productively detect cancer. Chinese researchers developed Deep Gene, an in-depth information and somatic point mutation classifier for cancer. Cancer can also be identified by extracting features from genetic data using a detailed learning approach. In contrast, in the diagnosis of cancer the complication Neural Network (CNN) is applied.

9.4.6 Treatment tailored

Modified treatment machine learning is a hot research issue. Its purpose is to provide better service with predictive analysis based on personal health data. The design of a revised therapy model based on symptoms and genetic information is based on mathematical and numerical machine learning methods (Fig. 9.11).

A supervised machine learning algorithm is used to build the modified treatment system. This method is established in series by means of patient medicine. An example of modified care is the Skin Vision device. This app allows you to check your skin for cancer on your mobile. The changed method of treatment will reduce healthcare costs.

9.4 Machine learning use in health care **189**

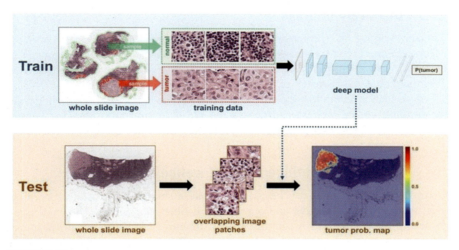

FIGURE 9.10

Detection and prediction of cancer.

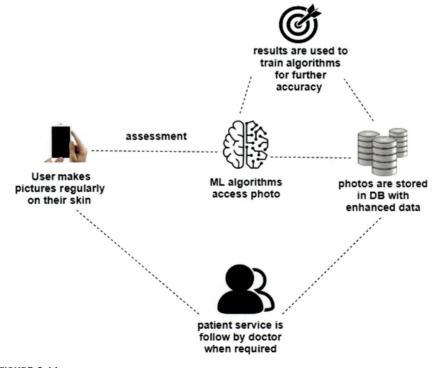

FIGURE 9.11

Image accessing by skin vision device.

9.4.7 Discovery of drugs

Machine learning is a common use of machine learning in medicine recognition. Microsoft Project Hanover aims to provide precision medicine for machine learning technologies. Many companies currently apply drug discovery machine learning techniques. Their purpose in the discovery of medicines is to use artificial intelligence (Fig. 9.12).

More than a few dollars are being reimbursed for these rural typewriters, such as speeding up the process and decreasing the deception rate. Machine learning also optimizes the development process and drug discovery costs.

9.4.8 Recorder of intelligent digital wellbeing

Computer learner ranges can be used to create a smart digital health record system, such as data cataloging and optical moral fiber recognition. The challenge is to develop a system which can sort patient inquiries by e-mail or turn a manual process into a computerized system. This software aims to create a platform that is safe and readily available.

The rapid expansion of electronic health records has enriched the medical information collected on patients for humanization in healthcare. Data errors are minimized, double data are reduced for design (Fig. 9.13).

Supervised machine learning algorithms, such as support vector machine (SVM), can also be used as a classifier for creating the electronic health recorders, or an Artificial Neural Network (ANN) can also be useful.

9.4.9 Radiology machine learning

Researchers have been working to combine radiological device learning and artificial intelligence. Aidoc provides the radiologist with tools to speed up the finding process

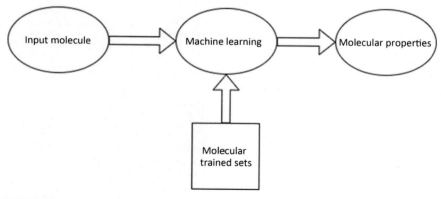

FIGURE 9.12

Drug discovery.

9.4 Machine learning use in health care

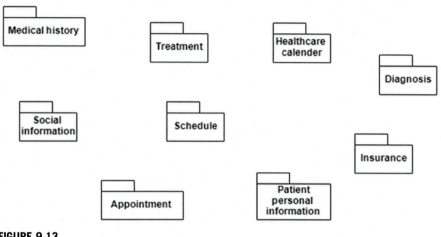

FIGURE 9.13

Electronic medical record.

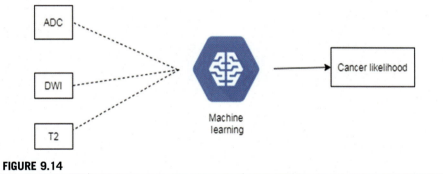

FIGURE 9.14

Radiology machine learning.

using computer teaching techniques. Their task is to analyze the clinical image in order to offer a simple explication for the diagonal classification of anomalies. This is usually the supervised machine learning algorithm (Fig. 9.14).

The machine learning approach is used for clinical object segmentation. Segmentation is the mechanism by which structures in an object are established.

9.4.10 Study and clinical trial

The clinical test can be a series of questions that require answers to determine a biomedical or pharmaceutical person's efficiency and safety. This test is aimed at focusing on new treatment development.

192 **CHAPTER 9** Cancer prediction and diagnosis hinged on HCML

The clinical trial is very time-consuming. Machine learning has a major impact in this area. An ML platform can provide reliable operation and real-time monitoring.

Machine learning methods can be controlled remotely in clinical trials and science. Machine learning also provides patients with a healthy clinical environment. The reliability of the clinical trial can be improved by supervised machine learning in health care.

9.5 **Cancer in healthcare**

Announcement of cancer patients is always incorporated into a complex set of interrelated components, i.e., a process. Patients with cancer are particularly vulnerable to a system of health care. Cancer and diagnosis can have major physical and emotional consequences. The good news is that the various people making up your healthcare team are willing to provide support. It may seem simple, but you must know from the heart that you are the key part of your healthcare team. You are a consumer of services and you shouldn't be afraid to ask questions about who and what you receive on condition that you get every type of health care that you get.

Adequate controls can reduce mortality and morbidity from colorectal, breast, and cervical cancer. Nevertheless, the full benefits of screening will be accomplished when effective application strategies are needed. They have carried out a systematic review to determine approaches for increasing the level of screening, breast, cervical, and colorectal (CRC) cancer as part of a broader implementation plan.

Interventions included customer warnings, product incentives, mass media, local press, group support, one-on-one practice, removal of corporate obstacles, and reduction of out-of-pocket costs. Our key finding, the success of the trials, was calculated as an improvement to the overall mean absolute percentage point (PP) of the average postintervention tests.

9.5.1 **Methods**

Our first step was an iterative analysis of the research background. This resulted in three systematic ratings of high quality. We have done a systematic analysis as our evidence base. In MEDLINE, EMBASE, and PSYCHinfo, randomized controlled trials released between 2004 and 2010 were scanned.

9.5.2 **Result**

Sixty-six experiments with 74 correlations created qualified studies. The reliability of the new studies was important. Consumer returns, limited press, and review and input from suppliers tend to be effective measures for increasing the uptake of screening in three cancers. Structural obstacles are also being taught one-on-one and minimized, but the role they play in CRC and cervical screening is less well-known. Further research is required in order to evaluate customer incentives, mass media, training groups, cutting baggage costs and incentives for suppliers.

9.6 Breast cancer in IoHTML

9.6.1 Study of breast cancer using the adaptive voting algorithm

Training in machines is a postulation of artificial intelligence (AI), which controls the facility to study and evolve automatically with no specific programming. Machine learning concentrates on developing information-inducing computer programs. The fundamental premise of machine learning is to use input information algorithms to predict an outcome when new data are available, using statistical research. Health care maintains and restores health, especially by trained and qualified professionals (as in medicines, dentistry and public health), by treating and preventing diseases. The HCML is useful because it helps the medical practitioner in the planning and distribution of large data sets away from the range of human capacity, and then efficiently transforms analyses of this information into clinical knowledge, which eventually results in better results. Machine learning applications in health care include identifying and diagnosing diseases (DD), drug discovery and manufacturing (DD&M), medical imaging diagnostic (MID), component behavior change in machine learning, intelligent records of health, better radiotherapy, and outbreak prediction. Breast cancer is one of the world's most highly active types of diseases for men and women. According to medical experts, this cancer helps save lives in its first phase.

Breast Cancer BC Adaptive Voting Ensemble Diagnostics HCML.

A large number of women pass away each year of cancer of the breast. It takes a considerable amount of time to diagnose breast cancer manually and it is hard for doctors to test. It is therefore very important to detect cancer through various automated diagnostic techniques. For breast cancer prediction and diagnosis, machine learning algorithms are available. Naïve Bayes (NB), Support Vector Machine (SVM), K-Nearest Neighbor (KNN) are some of the machine-learning algorithms.

9.6.2 Software development life cycle (SDLC)

The Lifecycle for Programming Improvement is an orderly procedure for programming building that guarantees programming quality and rightness (Fig. 9.15). The objective of the SDLC framework is to create great programming that meets the client's desires. Programming advancement ought to be finished inside the course of events and costs are predefined. Here are the essential reasons why a product framework is critical to SDLC: it provides a system for institutionalized exercises and accomplishments; it gives a component to extend observing and checking to build the perceivability of venture making

FIGURE 9.15

Software development life cycle (SDLC).

FIGURE 9.16

Work commitment.

arrangements for every one of the partners associated with the improvement procedure; improved client connections; and upgrades and improves advancement speed.

9.6.3 Parts of undertaking duty PDR and PER

The iterative life cycle portrays two significant capacities in which contemplations and qualities are shared firmly together the network of end clients and the advancement group (Fig. 9.16).

The point of the procedure is to characterize an approach dependent on the idea of "pair programming," which for a product venture is mainstream to one developer and one end user While it is hard to set up close connections between the different end-client network individuals and the product advancement group, it is a lot simpler to set up cozy connections between the pioneers. On the off chance that a few end-clients are associated with a few individuals from the advancement group, there will be contact between the two gatherings.

As the quantity of members expands, it decays. The model ought to speak with the engineer group, if important, yet it is dependent upon all partners to keep the PER and PDR exceptional, for instance, empowering PER and PDR to determine clashes that emerge when two diverse end clients impart varying necessities in a similar application.

9.6.4 Info structure

Passage configuration is a significant piece of gadget structure. The fundamental target, as demonstrated as follows, is to create a financially savvy strategy for yield.

- To accomplish the most ideal level of exactness.
- Ensure that the client comprehends the info and acknowledges it.

9.6.5 Input stage

The main entry steps prior to the information being processed in the server media are recording, copying of data, data preprocessing, data transformation, data management, data analysis, verification of data, correction of data, etc.

9.6.6 Output design

Computer outputs are primarily required to communicate process results to users. They are also used to provide the results for further consultation on a permanent basis.

- External outgoings, the destination of which is external to the organization.
- Internal outputs, the purpose of which are coordinated and the main interface with a computer.
- Outputs external to the organization.
- Operational inputs that are used in the computer department alone.
- Outputs for applications that specifically interact with the user.

The findings had to be written and recorded on the computer. The output type from the inputs generated after hand processing is taken into consideration. The regular printer is used as a source media for hard copies.

9.6.7 Responsible developers overview

- Design of the SRS program and resolution of all the system requirements.
- Demonstrate the system and install the device in the customer's site after a positive acceptance test.
- Send the required user manual detailing the interfaces of the application to be managed and system documents.
- Perform any user training necessary to use the program.

The gadget will be kept up upon organization for a term of 1 year.

9.6.8 Data flow

1. Dataflow is a term used in computing that has different meanings depending upon the application and the context in which the term is employed. Dataflow relates to stream treatment or reactive programming in the context of software architecture.
2. DFD membership ensures that the same data comes from a data store or sink to a specific location from two or more different procedures.
3. A data stream cannot go back to the process that leads to it directly. In order to generate another data flow, an additional program that controls the data flow will return the initial data.
4. Updating (deleting or changing) the flow of an information into a data store.

9.6.9 Cancer prediction of data in different views

In therapeutic analysis, the forecast of an infection goes about as a significant center in breaking down the restorative pictures. The undesirable cell development in any piece of the organ is known as a tumor. The tumor might be amiable or harmful. A threatening tumor is viewed as the most risky tissue. Thus the early finding of the sickness averts the malignant growth. In ladies, breast malignancy is treated as the hugest issue. This paper plans to survey on different information mining methods that are explicitly considered on breast disease forecasts.

9.6.10 Cancer predication in use case view

(Fig. 9.17).

9.6.11 Cancer predication in activity view

Disease predication in Activity view is a graphical portrayal of work progression of step savvy movements and activity with the aid of verdict and cycle simultaneousnessly (Fig. 9.17). Malignant growth predication in Activity view shows by and large the progression of control (Fig. 9.18).

9.6.12 Cancer predication in class view

A class outline imagines connections between classes. Interclass connections incorporate associations, generalizations, collections, and conditions. The present class graph likewise affirmed a number of relationships. "Speculation" demonstrates a connection in which one of the related classes (super class) is more general compared with the other (subclass) (Fig. 9.19).

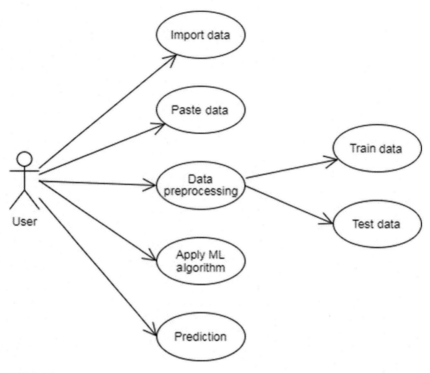

FIGURE 9.17

Cancer predication in use case view.

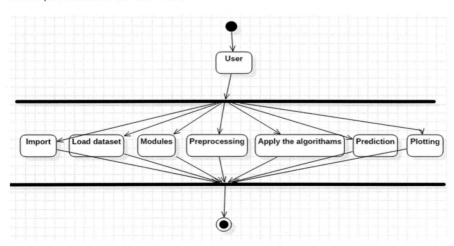

FIGURE 9.18

Cancer predication in activity view.

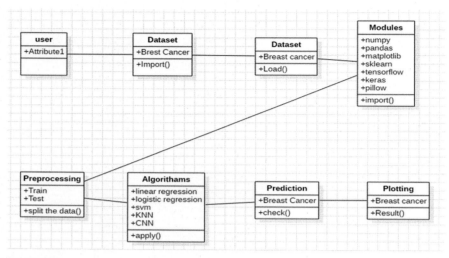

FIGURE 9.19

Cancer predication in class view.

9.6.13 Cancer predication in state chart view

The informational index is first imported and afterward we will stack the dataset and then import modules and process the informationThe calculations are applied for forecasting. At that point the information has been pictured by utilizing a few plots like pie-plot. At that point the model is prepared by utilizing random woods classifier and the forecast is done to know the precision of the model that has been constructed. At long last the outcome is shown (Fig. 9.20).

9.6.14 Symptoms of breast cancer

Cases of use identify the steps suggested by the object irrespective of its internal structure. Such interplays, which include connections between the actor and topic, may alter the state of the subject and connect with the audience.

- Scarring of the skin.
- A visible pull or indentation of the breast.
- Change of the nipple.
- Unusual release of the nipple.
- Change the feeling, size, or shape of the breast tissue.
- Scarring, lump, or hump of the leg.

9.6.15 Breast cancer types

While repeating malignancy spreads to different pieces of the breast from breast channels or organs, nonobtrusive disease doesn't spread from the tissues.

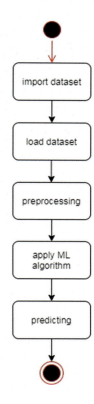

FIGURE 9.20

Cancer predication in state chart view.

In the breast malignant growth, it is ordinarily called "invasive" (obtrusive), "inside situ," or "intrusive."

Those two definitions are utilized to arrange breast malignancy's most basic sorts, including:

- **In situ, clinical carcinoma**: The nonintrusive condition is the ductal carcinoma in situ (DCIS). DCIS limits malignancy cells to the conduits in your breast and has not totally infiltrated your breast tissue.
- **Lobular in situ carcinoma**: Lobular in situ (LCIS) carcinoma is a malignancy that happens in your breast milk organs. In contrast to DCIS, the encompassing tissue has not been penetrated by malignancy cells.
- **Intrusive diarrheal carcinoma, invasive ductal carcinoma (IDC)** are the most widely recognized kind of breast malignant growth. This kind of breast malignancy begins in the milk feed of your breast and afterward drops into the surrounding tissue. In the event that breast disease has spread to the tissue outside your milk ducts, it might spread to other surrounding tissues and organs.

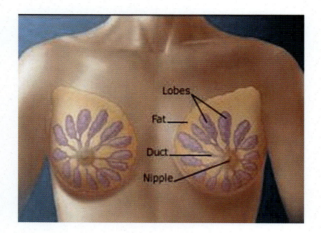

FIGURE 9.21

Breast cancer illustration.

- **Obtrusive carcinoma of the lobules**: Intrusive lobular carcinoma (ILC) grows first and attacks the surrounding tissue in your breast lobules. Other, less basic breast malignancy types include:
- **Areola paget infection**: This type of breast malignant growth begins in the areola channels, however, as it builds up, the skin and the areola change.
- **Tumor phyllodes**: In the connective tissue of the breast is this exceptionally uncommon sort of breast malignancy. Most of the tumors are benign, yet some are cancer-causing.
- **Agnihotra**: It is a malignancy which on the veins or on the lymph vessels.

You survey your treatments and the conceivable long-term results by the kind of malignancy that you have. Presently, AI techniques are utilized to deliberately find and distinguish tumors. For malignant growth finding, significant reasoning likewise assumes a significant job. Profound examination is anything but difficult to access and information sources can be obtained. An exploration has indicated that profound learning diminishes the danger of mistakes on determining breast malignant growths (Fig. 9.21).

Machine learning has demonstrated its ability to productively detect cancer. Chinese scientists investigated DeepGene: a classifier of cancer type using in-depth knowledge and somatic point mutations. Cancer can also be identified by extracting characteristics from genetic appearance information using a deep learning approach. In contrast in cancer diagnosis, the Complex Neural Network (CNN) is used.

9.7 Case study in breast cancer

This case reflects the new postmenopausal, HR-positive, HER2-negative, lower breast cancer treatment model that was not historically treated.

9.7.1 History and assessment of patients

A 65-year-old woman, with no previous medical record for breast cancer, named as a newly discovered oncology breast cancer surgery center. It developed pressure in her breast and a visible breast mass three months ago.

In both the mammogram and ultrasound, link breast weight was observed with the large ipsilateral lymph node. Invasive ductal carcinoma and a 95% positive estrogen receptor (ER), 85% positive, and a negative HER2 were found in the mass core biopsy (Fig. 9.22).

Adencarcinoma was a good target for the outstanding auxiliary lymph node needles. Tomography/computed tomography of positron emission was obtained and several 1–2 cm pulmonary nodules and expanded meditational and hilar lymph nodes were reported. For a core biopsy of one of the pulmonary nodules, interventional radiology was consulted. Metastatic breast cancer, 95% positive ER, 90% positive progesterone, and the negative HER2 were confirmed by biopsy.

9.7.2 Recommendations for diagnosis

A mix of Letrrozole 2.5 mg/mouth/day with Palbociclib 125 mg/mouth was allowed for 21 days during this underlying therapeutic oncology visit, followed by a 7-day rest period (for the fulfillment of a 28-day cycle). The program was bolstered by the National Complete Malignancy System direction for postmenopausal ladies with metastatic breast disease.

The main line postmenopausal HR-positive HER2 negative early breast abrogation and the letrozole + fake treatment test were tried in a stage II randomized preliminary (Dad LOMA-1 with letrozole in addition to palbociclib). The palbociclib results (20.2 months versus 10.2 months) show a noteworthy increase in

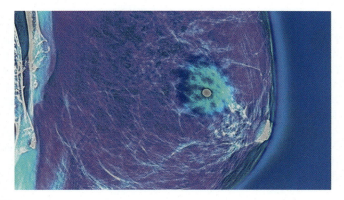

FIGURE 9.22

Cancer detection.

movement-free (PFS) endurance. There was a propensity in this network for better general endurance, yet it was not critical.

The doctor came to the conclusion of metastatic breast malignancy before recommending letrozole in addition to palbociclib and that the point of treatment is to reverse the movement of the disease, improve personal satisfaction, and increase endurance. Metastatic breast tissue is considered to be here and there not only relaxed but also treated as the mainstay of combination therapy.

The patient has addressed common adverse events such as joint pains, hot flushes, and increased risk of osteoporosis. The prescribed measures to protect against osteoporosis and improve general health were sufficient vitamin D and supplementation of calcium and regular training.

The most widely recognized negative manifestations are neutropenia, leucopenia, and weariness. The prescription is likewise connected with an expanded danger of venous thromboembolism (1%−5% of cured patients). Such potential negative impacts have been accounted for in the patient and a customary blood check is required (at regular intervals for the initial two cycles and afterward before each cycle).

9.7.3 Discourse

About 1.5 million new cases, 60% of which are HR-positive, are recorded every year around the world. Treatment is essentially performed by hormone treatment to treat HR-positive, dynamic HER2-negative breast cancer. Resistance to hormone treatment progressively builds, which prompted the growing enthusiasm for tweaking pathways for opposition. Cell cycle administrative disturbances are a potential system of hormonal treatment obstruction that is managed by the precise advancement of the cell through a gathering of proteins which incorporate CDKs (Fig. 9.23).

The improvement of other 4−6 CDK inhibitors, including ribociclib and abemeclib, likewise gives positive outcomes. Consequences of random, double-blind visual acuity, multifocus, replanted I, II, randomized stage III preliminary of 668 postmenopausal ladies with HR-positive, HER2-negative propelled breast malignant growth not recently treated for cutting edge breast disease were randomized to ribociclib (400 mg day by day, after 3 weeks, and multi week off) and fake treatment related to Letrozole 2.5 mg every day.

MONALEESA-3 is a continuous research preliminary for ribociclib with fulvestrant relative with fulvestrant alone for people who experience the ill effects of HR-constructive, HER2-pessimistic breast disease with no or no past line of endocrine treatment. This is a Stage II study for endocrine-safe patients who were broadly pretreated for cutting edge sickness. A middle PFS of just about a half year initiated Abemaciclib in those patients by a reaction pace of right around 20%.

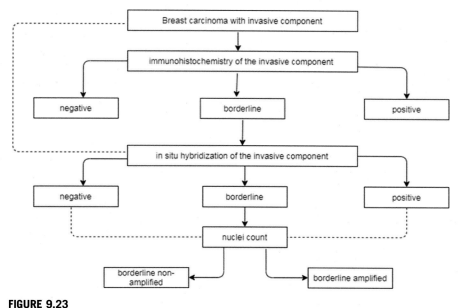

FIGURE 9.23

Cancer detection.

9.7.4 Outcomes of diagnosis

Of total neutrophil counts (ANC) of 1200/mm^3 (grade 2 neutropenia), follow-up of a complete blood count 4 weeks after starting treatment was excellent. The woman felt good and she was feverish, so she started with palbociclib. ANC fell to 800/mm^3 4 weeks later and the patient was still affebrile. Since this was grade 3 neutropenia, palbociclib was held for 1 week and was resumed when ANC was >1000/mm^3. In the absence of other adverse effects, she well accepted the medication. The simulated head, stomach, and pelvis tomography scannings were conducted 3 months after starting treatment. Brain mass volume, lymph nodes, and lung nodules decreased markedly. The primary line of treatment was additionally the blend of ribociclib and letrozole with cutting edge HR-positive, HER2-negative breast disease in the huge postmenopausal ladies, and PFS demonstrated huge advancement dependent on a fundamental, prearranged break investigation.

A subsequent calculation in breast malignant growth patients was dependent on chance class (Fig. 9.24).

A therapy has grown potentially through acceptance of endocrine opposition, for cutting-edge, HR-positive, HER2-negative breast malignant growth. CDK 4/6 inhibitors, including palbociclib, rival and cleavage. Palbociclib plus letrozole was verified with first-line findings for advanced postmenopause in PALOMA-1 tests showing a considerably improved PFS in the palbociclib family.

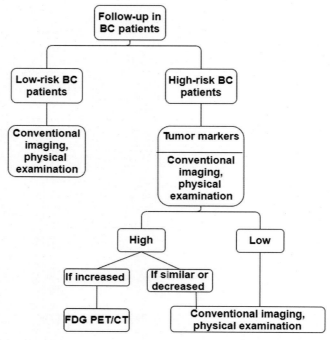

FIGURE 9.24

Subsequent calculation.

9.8 Breast cancer algorithm

Naive Bayes classifiers are known for their development of simple but well performing models, particularly in the field of the classification and disease prediction of documents, and are based on a common theorem of probability in Bayes. They will look at the principle of naive Bayes classifications in the earliest part of a series and introduce the fundamental concepts of text classification. We will use the following articles to construct a naive Bayes spam filter and apply naive Bayes to lyric-based song classification.

Scientists have taken this problem very seriously for over half a century ago: "Should we create a model that automatically learns for the available data and makes the right decisions and expectations?" This sounds almost a rhetorical question, and the answer to it is posed in various applications including pattern recognition, machine learning, and artificial intelligence.

Data from different sensor devices in combination with strong learning algorithms and domain knowledge have led to numerous great inventions that we now regard as obvious in our daily lives: internet inquiries via search engines like Google; text recognition at the post office; supermarket barcode scanners; disease diagnosis; voice recognition on our mobile phones.

One subfield of prediction modeling is the supervised identification of patterns. It is important to train the system on the basis of identified learning information to apply a predefined class tag to new objects. An example is spam filtering using Bayes Classifiers to determine whether or not a new text message can be marked as spam. Naive Bayes Classifiers are known for creating simple but effective models, especially in the fields of media identification and disease prediction, based on the well-known Bayes theorem on possibility (Fig. 9.25).

Innocent Sound Classifiers are named as straight or proficient classifiers. The naïve Bayes characterization model depends on the Bayes hypothesis and the descriptive word is viewed as commonly independent. In small example sizes, in general, naïve Bayes classifiers can outperform powerful yet well-functioning alternatives [1], especially in reporting classification and disease prediction.

Bayes grading systems are relatively robust, easy, quick, and reliable in several fields. Examples include the detection of disease and decision-making in care [2], identification in taxonomic studies of RNA sequences [3] and filtering spam on e-mail clients [4].

Clear breaches of autonomy and nonlinear classification problems could lead to very poor performance in naïve Bayes classifications.

We must bear in mind that what kind of classification model we want to use depends on the type of data and the type of question. In theory, it is always advised to compare different classification models on the same data set and to consider both the approximation and the utility of the measurement (Fig. 9.26).

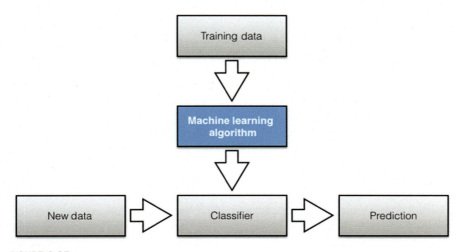

FIGURE 9.25

Classification.

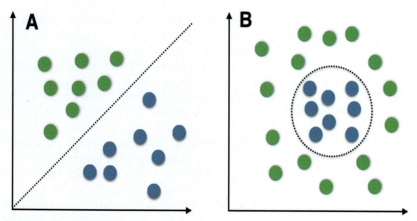

FIGURE 9.26

Classification model. Linear (A) versus nonlinear (B) issues. Random samples are shown as colored spheres for two different classes, and the dotted lines indicate the class borders which the classifiers attempt to approximate by the calculation. A non-linear problem (B) would not be ideal for linear classifiers, like naive Bays, because classes are not distinct linearly. Nonlinear classifiers (e.g., closest proxy classifiers) should be favored in such a situation.

9.9 Conclusion

Planning for resilience and disease treatment is part of the definition. Its principle intention is to fix malignancy patients or draw out their life remarkably, guaranteeing a civilized personal fulfillment. All collectively for a conclusion and treatment program to be powerful, it must never be produced in seclusion. It should be associated to a premature position program with the goal that cases are recognized at a beginning time, when treatment is progressively possible and here is an additional prominent possibility of fix. It furthermore should be included with a palliative deliberation program, so patients with callous edge malignancies, who can never again yield by treatment, will obtain suitable alleviation from their physical, psychosocial and otherworldly torment. In addition, projects ought to integrate a mindfulness raising part, to instruct patients, family and network individuals about the malignant growth hazard factors and the requirement for taking preventive measures to abstain from creating disease. Wherever possessions are constrained, conclusion and treatment administrations have to at first focus on all patients giving reparable tumors, for exemplar, breast, cervical and oral malignant growths that can be distinguished early. They could likewise incorporate youth intense lymphatic leukemia, which has a high potential for fix in spite of the fact that it can't be recognized early. Most importantly, administrations should be given in a fair and supportable way. As and when more assets become accessible, the program can be stretched out to incorporate other reparable diseases such as tumors for which treatment can draw out endurance impressively. As far as exactness, naïve Bayes have scored high figures of 0.9704 and 0.964 separately, without applying PCA. K-Neighbors (0.9349) and Logistic relapse (0.923) are not far behind either. SVM scores 0.917 in precision. Choice Tree plays out the most noticeably worst of the six at 0.834. Use of PCA decays the precision of the considerable number of calculations with the exception of the Decision tree. Be that as it may, the exactness figures are as yet higher than that of Decision tree's LDA, once more, performs best after PCA is applied, despite the fact that there is a fall in air conditioning curacy (0.917). Considering the other presentation network into account, a great deal can be resolved with respect to the exhibition of the calculations. Choice tree, K-Neighbors and Naïve Bayes per-frames better without the presentation of PCA, while LDA, Logistic Regression and SVM per-shapes better after PCA is applied to the dataset. SVM and Logistic Regression scores an ideal 1.000 with regards to review, which is essential as far as infection expectation, after PCA is applied, despite the fact that there are decreases in the estimations of all other exhibition measurements of both the mentioned calculations. Remembering that PCA decreases the run time exponential to immense reaches out in datasets (both little and huge the same) and keeping the review score into thought, we can presume that Logistic Regression and Support Vector Analysis with PCA performs better with regards to Breast Cancer Prediction for this dataset used.

References

[1] J.G. Elmore, C.K. Wells, C.H. Lee, D.H. Howard, A.R. Feinstein, Variability in radiologists' interpretations of mammograms, N. Engl. J. Med. 331 (22) (1994) 1493−1499.

[2] N. Mao, P. Yin, Q. Wang, et al., Added value of radiomics on mammography for breast cancer diagnosis: a feasibility study, J. Am. Coll. Radiol. 16 (4) (2019) 485−491.

[3] H. Wang, J. Feng, Q. Bu, et al., Breast mass detection in digital mammogram based on gestalt psychology, J. Healthc. Eng. 2018 (2018) 13. Article ID 4015613.

[4] A.J. Cruz, D.S. Wishart, Applications of machine learning in cancer prediction and prognosis, Cancer Inform. 2 (2006) 59−77.

CHAPTER 10

Parameterization techniques for automatic speech recognition system

Gaurav Aggarwal[1], Sarada Prasad Gochhayat[2] and Latika Singh[3]

[1]*School of Computing and Information Technology, Manipal University Jaipur, Jaipur, India*
[2]*Virginia Modeling, Analysis and Simulation Centre, Simulation and Visualization Engineering, Old Dominion University, Suffolk, VA, United States*
[3]*Ansal University, Gurugram, India*

10.1 Introduction

Speech plays a vital role in social and verbal communication for humans. "Verbal communication" means expressing ideas, feelings, and thoughts, and "speech" is a medium to deliver information, which is not restricted to statement content, but further extends to information about speaker's gender, age and purpose, and several other factors. The main components for a fluent speech are voice, articulation, and fluency. Voice is a sound produced by vocal cords and breathing; articulation refers to the capability to produce correct sound; fluency is the tempo of speech, such as how to deliver a fluent speech. These three components are interrelated and play a vital role in the production of an eloquent speech. With so many modes of communication, speech is hard to substitute. Anatomically, speech is articulated by the movement of the larynx that regulates the activities of vocal folds.

Speech utterances always have some disturbance, which causes speech disfluency in communication. It can be pathological or healthy. A pathological voice may comprise particular desertion between brain and speech motor control such as improper functioning of larynx muscles, causing typical speech severe to understand. It is recorded that a small quantity of blood flow and electrical signals in the brain participate in speech production.

Initial cognitive development is directly dependent on communication capabilities. Neurodevelopment disorders such as intellectual disability, autism, and Down's syndrome can directly affect speech and language development at initial and future stages. Intellectual disability (ID) is also a type of neurodevelopmental disorder which causes speech and language impairments, abnormal social behavior, and less interest in communication. It is somewhat an obsessive behavior.

According to DSN-V [1], intellectual disability is classified by substantial limitations in normal mental abilities and adaptive behavior that begin through the

Machine Learning and the Internet of Medical Things in Healthcare. DOI: https://doi.org/10.1016/B978-0-12-821229-5.00010-0
Copyright © 2021 Elsevier Inc. All rights reserved.

path of a child's development. A noticeable difference must be there when we compare a child with intellectual disability from a typically developed (TD) child of the same age, gender, and socioeconomic and cultural background. People have many issues in learning things quickly and efficiently, trying to think creatively and process new information.

Intellectual disability (ID) is a frequent and severe neurodevelopmental disorder which affects speech and language production. The ID children with speech impairment constitute 7% of early age population and the risk rises in premature infants [2]. The impairment is usually defined as an inability to attain standard communication skills with age. With sufficient peripheral learning, a child may be intelligent, and the lack of cognitive sensomotoric is not observed [3]. These diseases adversely affect the child's growth and personality. In this work, we deal with the techniques that can assess the development of the disease and try to find out what makes it difficult and may stop the children from learning to utter. Speech sentences spoken by intellectually disabled children are different from speech pronounced by typically developed children of the same age [4,5]. A professional speech pathologist can readily observe the difference. The pathologist is also proficient at determining the condition of the disease (mild to profound). Our goal is to develop a software tool that can support the pathologists and therapists in the process of predicting the disease at an early age. Since intellectual disability affects the children's speech capabilities, the classification of speech sentences helps in deciding whether a child is intellectually disabled or not, so that the appropriate treatment will be given to the child at an early age. The classification model based on analysis of speech should be capable of determining the category of the disorder: mild, moderate, severe, and profound. A relation has been observed between intellectual disability and the impairment of the speech utterances [6]. A large percentage of the population suffers from intellectual disabilities. Such a vast prevalence puts this disorder into the pool of the most frequently appearing neurodevelopmental disease that harms children [7]. Since the speech sentences are coded into movements of the vocal tract [8], it is possible to classify the disorder by supervised machine learning techniques [9]. However, this approach requires high and fast computational power and enabled hardware. But the software part is quite convenient and cheap enough to apply in the clinical practice.

10.2 Motivation

These days human—machine interaction has paved its way to new enhancements, and much effort is being made to build a system that is as easy to interact with as human interaction with each other using speech. Also, the system is intelligent enough to classify between ID and TD children using their voice correctly. The following questions can better explain the actual motivation that lies behind fulfilment of this work. What is the reason for classifying children into ID and TD? What are the extracted features from speech, so that the performance of the system is best? What are the significant steps taken during the preprocessing step to retain the performance of the system when used in the real world with noisy data? How and on which platform are the feature extraction techniques implemented? What are the classification algorithms used? What tool will be used for classification?

The motivation of the work is to implement an integral and robust system for the prediction and classification of neurodevelopmental disorders using their speech. This chapter aims to develop a system for speech application that is robust when used in the real world and with noisy data. For achieving this goal, a combination of speech signal processing and machine learning will be used. The study of speech signal processing covers various speech features and their respective feature extraction techniques, whereas the machine learning approach contains algorithms for the categorization of features extracted to predict the class. The features extracted are the most useful features for a speech recognition system. Four classification algorithms are used for the best classification accuracies. Apart from this, other advantages include a smooth implementation of the system. Also, it consists of speech of the user that is easily assessable.

10.3 Speech production

In humans, the natural phenomenon of speech production takes place through vocal cords in the glottis. The primary factor that produces speech is "vibration of the vocal cords." The major components in speech production include lungs, windpipe or trachea, larynx, pharyngeal cavity, oral cavity, and nasal cavity. The vocal cords are contained within the larynx, which is also known as the glottis.

The pharyngeal cavity comprises the throat, which in combination with the oral cavity, containing the mouth, forms the vocal tract. The nasal cavity is used for the production of nasal sounds. The nasal cavity is combined acoustically with the vocal tract when the velum is lowered. Fig. 10.1 depicts the sagittal view of the human speech production system.

The above-described organs are used for the production of sound through vocal cords; that is, they convert air into sound. The organs like lips, tongue, jaw, and teeth are used to give shape to the sound produced or the correct word or phoneme produced by the human. These organs are called articulatory organs or articulators. There are two types of articulators, namely, active articulators and passive articulators. The active articulators move, but the passive articulators do not. The former include lips and tongue, and the latter comprise upper teeth and upper jaw. Depending upon the position of these articulators different sounds are produced [10]. Fig. 10.2 shows the working of the human speech production system.

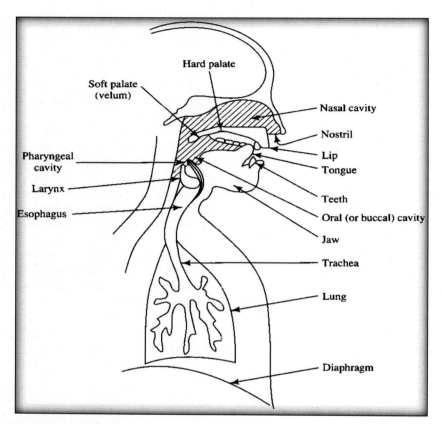

FIGURE 10.1

Sagittal view of human speech production system.

The system works as follows: the air expelled from the lungs passes through the windpipe and comes into contact with the vocal cords within the larynx (this phenomenon is also called input excitation), the vocal cords vibrate. It makes the airflow split into quasiperiodic waves which are further modulated by the articulators.

Speech simplifies the means of communication among humans. However, with the rapid advancement in the field of information technology, the use of computer-based applications has increased. Many innovations have evolved in recent years. One such change is the speech pattern recognition system. This research also aims to build a medium that allows the user to interact with the computer system in a way that is similar to human speech. The speech production in the machine is analogical to the human speech production system. Fig. 10.3 depicts the speech production by the computer and also explains the voiced and unvoiced states.

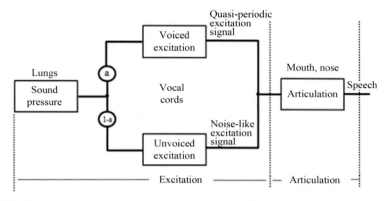

FIGURE 10.2

Human speech production system.

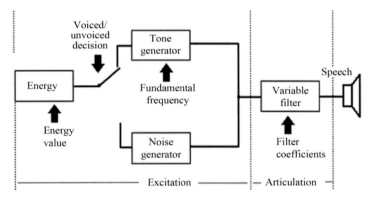

FIGURE 10.3

Speech production by machine.

Analogical to the human speech production system, the machine speech production system comprises two phenomena, namely, excitation and articulation. The energy is supplied to the system whose value decides the voiced and the unvoiced state. The decision was made by comparing the energy to some threshold value, the energy that is below the threshold value is detected as unvoiced, and the one that is above the threshold value is voiced. Just like vibration of the vocal cords, the value of fundamental frequency is given as the input to the tone generator that produces sound, which is then passed to the filters which decide the shape of the sound produced.

10.4 Data collection

The dataset was created from an organization named SIRTAR at Rohtak—owned by the government of Haryana, India. The dataset comprised recordings for monologues, picture naming, reading, and repeating sentences. There were 48 different recordings taken under analysis. The dataset consisted of 24 speech samples of children with ID (mild and moderate) and 24 speech samples of TD children. The distribution of the participants is shown in Table 10.1.

The content of the speech recordings were vowels, pictures of animals, fruits, and vegetables, and imitation of two letters words of both Hindi and English languages. The participants for the TD group were chosen from a nearby school. All the typically developed children were mentally and physically fit. Signed consent forms were taken from the research participants before involvement in the research. This research was approved by the ethical agency of SIRTAR, Rohtak, India.

10.4.1 Recording procedure

Speech samples were recorded at the sampling rate of 22,500 Hz with 16 bit pulse-code modulation (PCM). The children's voices were recorded with a head-mounted unidirectional Sony microphone (ICD-UX533F) with a dynamic gain control feature. The participant's instructor was present in the room while recording the speech, to make the research participant relaxed. Repetition was there to avoid the hesitation in the participant's voice. Three tasks were performed to create the speech dataset.

1. Sustained phonation task: research contributors were asked to speak Hindi and English vowels and alphabet and to count for 60 seconds.
2. Imitation task: research participants were asked to repeat the scholar utterances which largely consisted of two to three letters words of Hindi and English language. This task was performed to examine speed and articulatory correctness.
3. Picture recognition task: a photorecognition task comprised pictures of standard colors, vegetables, birds, and animals. The participants were asked to recognize and pronounce the name of the picture showed. This task records various dimensions of voice quality, speaking rate, and voice intensity with duration of pauses, duration of spoken syllable, and sentence and voice onset time.

Table 10.1 Research dataset.

Type	Age (mean)	Gender	
		Male	Female
ID mild and moderate	13.9 years	18	6
Controls	14.2 years	16	8

10.5 Speech signal processing   **215**

Proper training was given to the participants before the actual recording procedure to avoid any sort of psychological stimuli in the children's speech.

10.4.2 Noise reduction

As the utterances were recorded in the closed room of the school, the breathing and thumping sound, adjustments of the mike, chair movements, and many unwanted sounds were recorded in the speech signal. For this reason, the processing was done using the method capable enough to neglect or remove these disturbances. For medical applications, Goldwave, a speech processing tool for preprocessing and analyzing the speech datasets was used. For noise reduction, Goldwave Noise Reduction Technique was applied for reducing the noise from the signal by using a frequency analysis technique. The process to reduce the noise is as follows: the mp3 signal was uploaded from the computer into the software and the maximize volume function was used to increase the signal volume to 100%. Then selection of the noise reduction filter function reduced the noise from the speech signal. The overlapping window was set at $16 \times$ FFT size 12 and 90% scaling. To eliminate the tiny artifacts the scaling window was taken up to 90%, Using these settings a noise free speech signal was prepared which can be used for speech parameterization.

Recorded utterances were authenticated with the cooperation of speech pathologists and psychologists. It included different types of single phoneme and complete sentences. The classification task was time-consuming due to a large number of features in the dataset. Therefore a feature selection algorithm (filter method) was applied to choose the best features from the whole set.

10.5 Speech signal processing

A speech signal comprises a vast variety of essential features containing information about the gender and emotion of the speaker, and voice quality, linguistic clarity, and pathologies present in the speaker's sound. Several speech parameters are directly calculated from the acoustic signal, while some are inferred, like speech pathologies due to neurodevelopmental disorders. Regardless of significant research in the field of speech analysis, speech pattern recognition still uses the inference of machine learning. The block diagram of feature extraction is shown in Fig. 10.4. Further, we will elaborate each step in detail.

In signal processing, a signal is nothing but an experimental measurement of any event [11]. The upward speed of a rocket, the value of the bitcoin, an infant's cry are examples of signals in various fields. A signal is represented as a function of some independent variable, usually time, and it is denoted as f(t). A signal can be a function of one or more variables. An image signal is a function of two variables f(x;y) which signifies the color of a pixel (x,y).

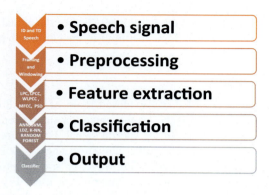

FIGURE 10.4

Block diagram for feature extraction.

There are two types of signals: analog and digital. A signal with continuous domain and range is called an analog signal. Analog signals are difficult to store on a computer for processing. Therefore an analog signal is first converted to a digital signal and then analyzed on a computer system, where the ranges and the domains are discrete.

10.5.1 Sampling and quantization

An analog to digital (A/D) converter is used to digitize an analog signal. First, the signal strength can be measured at the particular points. This process is called sampling. To elaborate, let $x_a(t)$ be an analog signal as a time function t. On sampling of x_a with a sampling period T, the digitized output of this procedure is $x[n] = x_a(nT)$. The inverse sampling period Fs = 1/T is interpreted as the sampling frequency F_s. Fig. 10.5 describes the sampling process of a sinusoidal signal.

Second, the sampled readings of the signal must be transformed into a discrete set of values. This technique is known as quantization. In speech samples, the quantization level is usually set as the number of bits required to define the range of the signal. The range of a 16-bit signal is from −32,768 to +32,767. Fig. 10.6 shows the quantized analog signals. These processes cause the information losses of a signal. This creates noise and error in the output signal. The appropriate method of avoiding these noises is to speed up the sampling process to adequately rebuild the original signal. In the process of quantization, the big issue is to choose between the size and the output quality of a signal. Generally, a digital system is a process that receives information as an input signal and performs appropriate tasks in the form of an output signal as shown in Eq. (10.1).

$$y[n] = T\{x[n]\} \tag{10.1}$$

10.5 Speech signal processing

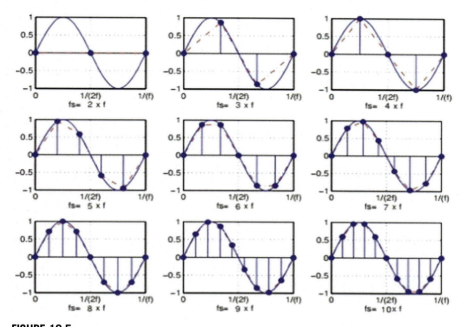

FIGURE 10.5

Sampling process of a sinusoidal signal [11].

10.5.2 Representation of the signal in time and frequency domain

Vibration in vocal cords is primarily responsible for speech production. This process generates sound pressure, which changes according to air pressure causing sound. The calculation of sound pressure is known as amplitude. This sound is depicted as a speech waveform in the time domain. It represents a change in amplitude with time. Fig. 10.7 represents a plot of a speech signal.

The signal envelope defines an oscillatory nature of a speech waveform, i.e., periodicity over a time period. It is clearly shown in Fig. 10.8. So an analog signal with period T is periodic with a condition:

$$x_a(t + T) = x_a(t) \ \forall t \tag{10.2}$$

Similarly, a digital signal $x[n]$ is periodic with period N if and only if:

$$x[n + N] = x[n] \ \forall n \tag{10.3}$$

From Eqs. (10.2) and (10.3), it is observed that a periodic signal must satisfy Eq. (10.2) (in case of an analog signal) and Eq. (10.3) (in case of a digital signal), otherwise, it is an aperiodic or nonperiodic signal.

Apart from the time domain, a signal can be viewed in the frequency domain also. The best example for this is Newton's prism experiment [14]. On passing a

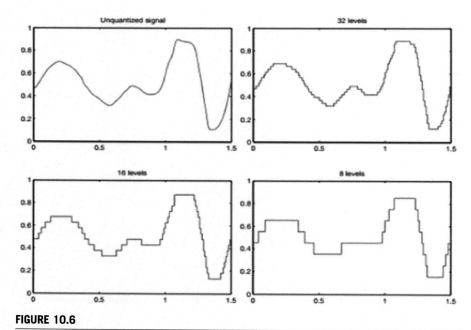

FIGURE 10.6

Quantization process of a sinusoidal signal [12].

white light from a prism, the light splits into a band of colors and when the band of colors passes through a second prism, thitey again reconstructs the white light. Decomposing a white light into the color spectrum is an analysis in the frequency domain.

In digital signal processing, the sinusoid signal, as shown in Eq. (10.4), plays an important role:

$$x_a(t) = A\cos(\omega t + \varnothing) \quad -\infty < t < \infty \qquad (10.4)$$

where A is the amplitude, ω is the angular frequency, and $\varnothing$ is the phase of the signal. The values of angular frequency and phase are in radians. The angular frequency is also related to frequency f, in hertzs with an Eq. (10.5):

$$\omega = 2\pi f \qquad (10.5)$$

The sine wave signal is periodic with period T = 1/f Eq. (10.2). The digitized form of this equation is represented as:

$$x[n] = A\cos(\omega n + \varnothing) \quad -\infty < n < \infty \qquad (10.6)$$

According to Eq. (10.3), x[n] is a periodic function with period N and a condition, i.e., $\omega = 2\pi N$ is a rational number. Hence, the Eq. (10.6) of a digital signal is aperiodic for all values of ω.

10.5 Speech signal processing   219

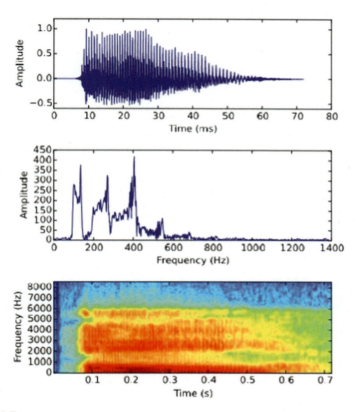

FIGURE 10.7

A male speech sample at 44 kHz (A) waveform (B) spectrum limited to 1400 Hz (C) spectrogram downsampled from 0 Hz to 8 kHz.

Frequency transformation is a process in which a signal is transformed into a frequency domain from the time domain. A spectrum represents the sound in a frequency domain as it plots the amplitude at its matching frequencies, whereas a spectrogram provides the three-dimensional information of a sound signal. The level of amplitude is represented by the shade/color of every time-frequency point. The amplitude is directly proportional to the darkness of the point (hotness of the color point). Spectrograms are proven to be a useful visual tool to study the speech acoustics.

10.5.3 Frequency analysis

Fourier analysis is a mathematical technique which is applied to transform a signal from the time domain to a frequency domain. Depending upon the nature and

220 CHAPTER 10 Parameterization techniques for automatic speech

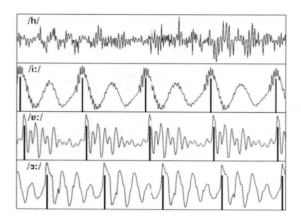

FIGURE 10.8

Periodic and aperiodic speech signal (from [13]). The waveform of voiceless fricative /h/ is aperiodic, while for the three vowels it is periodic.

Table 10.2 Types of fourier techniques.

Time domain properties	Periodic	Aperiodic	
Continuous	Fourier series (FS)	Fourier transform (FT)	Aperiodic
Discrete	Discrete fourier transform (DFT)	Discrete time fourier transform (DTFT)	Periodic
	Discrete	Continuous	Frequency domain properties

the periodicity of the signal one can choose the appropriate Fourier technique. Four types of Fourier techniques are shown in Table 10.2.

Each technique is a combination of two transformations. The Fourier Series (FS) of a periodic signal x(t) with period T is defined in Eq. (10.7) and (10.8):

$$c_k = \frac{1}{T} \int_T x(t) e^{-j2\pi kt/T} dt \qquad (10.7)$$

$$x(t) = \sum_{k=-\infty}^{\infty} c_k e^{j2\pi kt/T} \qquad (10.8)$$

The Fourier Transform (FT) of an aperiodic signal x(t) is defined in Eqs. (10.9) and (10.10).

$$X(\omega) = \int_{-\infty}^{\infty} x(t) e^{-j\omega t} dt \qquad (10.9)$$

$$x(t) = \int_{-\infty}^{\infty} X(\omega)e^{j\omega t}d\omega \qquad (10.10)$$

The Discrete Fourier Transform (DFT) of a discrete periodic acoustic signal x [n] with period Nis defined in Eqs. (10.11) and (10.12).

$$c_k = \frac{1}{N}\sum_{n=0}^{N-1} x[n]e^{-j2\pi kn/N} \qquad (10.11)$$

$$x[n] = \sum_{n=0}^{N-1} c_k e^{j2\pi kn/N} \qquad (10.12)$$

The Discrete Time Fourier Transform (DTFT) of a digitized aperiodic signal x [n] is defined in Eqs. (10.13) and (10.14).

$$X(\omega) = \sum_{n=-\infty}^{\infty} x[n]e^{-j\omega n} \qquad (10.13)$$

$$x[n] = \frac{1}{2\pi}\int_{2\pi} X(\omega)e^{j\omega n}d\omega \qquad (10.14)$$

10.5.4 Short time analysis

Speech signals are nonstationary in nature, means their statistical parameters like intensity and variance vary over time [11]. Speech signals may be stationary for a shorter period but when considered over a longer duration they are aperiodic. Therefore the Fourier transform is not a suitable technique for speech analysis as it requires a periodic signal for infinite time. A technique called short time analysis is used. In this technique, the signal is divided into short frames or segments, assuming the signal is stationary in that short frame and analyzing each frame or segment separately. The length of each frame is about $10-20$ ms, short enough to satisfy the assumption. The spectrogram is an example of short rime analysis which is discussed in the next section. Discrete-Time Fourier Transform is applied to each frame which results in spectra over time.

For a given signal $x[n]$, the short time signal $x_m[n]$ of frame m is represented by Eq. (10.15):

$$x_m[n] = x[n].w_m[n] \qquad (10.15)$$

where $w_m[n]$ is a window function, beyond a specific region its value is zero. For this technique, $w_m[n]$ should be equal for all frames. So the calculations are as in Eqs. (10.16) and (10.17).

$$w_m[n] = w[m-n] \qquad (10.16)$$

222 CHAPTER 10 Parameterization techniques for automatic speech

$$w[n] = \begin{cases} \hat{w}[n] & |n| \leq \dfrac{N}{2} \\ 0 & |n| > \dfrac{N}{2} \end{cases} \qquad (10.17)$$

where N is the length of the window.

10.5.5 Short-time fourier analysis

Using Fourier analysis of a signal x[n], from the Eq. (10.13), DTFT of the frame $x_m[n]$ is explained in Eq. (10.18).

$$X(m, \omega) = X_m(\omega) = \sum_{n=-\infty}^{\infty} x_m[n]e^{-i\omega n} = \sum_{n=-\infty}^{\infty} x[n]w[m-n]e^{-i\omega n} \qquad (10.18)$$

This Eq. (10.18) defines the short time DTFT for speech signal x[n].

10.5.6 Cepstral analysis

Bogert et al. [15] were the first to apply the word cepstrum as the inverse Fourier transform of the spectrum of a speech sample. This transformation was performed on a simple echo signal. The signal was separated into two parts: an original signal function and a periodic function of the echo frequency. The transformation was not the frequency of an independent variable. A new domain was invented as a frequency domain and the outcome of the procedure was called cepstrum.

In a previous study, the researcher worked on nonlinear signal processing as the notion of a homographic system [16]. In this system, the vector space of the input operation was mapped on the intermediate addition vector space, which was further mapped with the vector space of the output operation. The example of homomorphic transformation is described in Fig. 10.9. This application is called homographic filtering,

Consider homomorphic filtering with convolution as an input operation. The first component of the system is liable to map a convolution operation into a deconvolution or addition operation, which is shown in Eq. (10.19).

$$D(s_1(t) * s_2(t)) = D(s_1(t)) + D(s_2(t)) \qquad (10.19)$$

Also, this transformation can be achieved by the series combination of the Fourier transforms, logarithms, and inverse Fourier transform. The Eqs. (10.20−10.22) show the definition of the cepstrum of a discrete signal:

$$\hat{x}[n] = \frac{1}{2\pi} \int_{2\pi} \hat{X}(\omega)e^{j\omega n} d\omega \qquad (10.20)$$

where $X(\omega)$ is the DTFT of x[n] and:

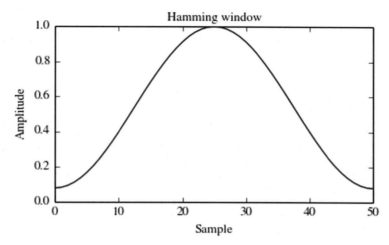

FIGURE 10.9

A 250-sample Hamming window function.

$$\hat{X}(\omega) = \log[X(\omega)] \tag{10.21}$$

Similarly, the real cepstrum is stated as Eq. (10.22).

$$\hat{x}[n] = \frac{1}{2\pi} \int_{2\pi} \log\big[|X(\omega)|\big] e^{j\omega n} d\omega \tag{10.22}$$

10.5.7 Preprocessing: the noise reduction technique

Generally, people from different backgrounds have significant variations in their speech characteristics. In recent years, speech pattern recognition has become a challenging task due to speakers with different socioeconomic backgrounds, a noisy or quiet environment, and their speaking abilities. Performance of speech pattern recognition reduces to a great extent when background noise is recorded with the speech signal. Among all the background noises, environmental noise is globally present, so it pulls the attention of the researchers. Many noise reduction techniques have been proposed to deal with such types of noises. The environmental noise consists of additional sounds produced from wind, machines, breathing, and tapping of hands and can be superimposed with the speech signal in the time domain to produce the perceived signal. The spectral subtraction method is used to cancel out the noise from the original signal by estimating average noise in nonspeech regions and then deducting it from the observed speech signal. There are many current applications such as restoration of musical recordings

224 CHAPTER 10 Parameterization techniques for automatic speech

[17,18], mobile telecommunication [19], and speech recognition [20,21] where we are applying the spectral subtraction method for noise reduction.

However, the speaker usually notices an increase in amplitudes and formant frequencies with high background noise. The phenomenon is called the Lombard effect [22]. A second noise reduction technique is mean and variance normalization (MVN) which is used to lessen the distortions of the speech in the frequency domain. Convolutional noise is another environmental variation which is produced from channel effects like impulse response of vocal tract, room, microphone, etc. A normalized technique, namely the cepstral mean subtraction (CMS) method, is generally used to minimize this type of noise [23]. The cepstral mean subtraction technique is discussed in the spectral subsection.

10.5.7.1 Spectral subtraction method

Preprocessing is performed on the original input signal to obtain a more reliable speech signal from the noisy signal x[n], as shown in Eq. (10.23).

$$x[n] = s[n] + d[n] \tag{10.23}$$

where s[n] is the noise free-part of the speech sample and d[n] is the additional background noise present in x[n]. In order to extract the speech features from the noisy signal x[n] in such a way that the extracted spectral features contains less noise d[n].

By taking the Fourier transform, we get Eq. (10.24).

$$\sum_{n=-\infty}^{\infty} x[n]e^{-j\omega n} = \sum_{n=-\infty}^{\infty} s[n]e^{-j\omega n} + \sum_{n=-\infty}^{\infty} d[n]e^{-j\omega n} \tag{10.24}$$

Giving

$$X(\omega) = S(\omega) + D(\omega) \tag{10.25}$$

After noise reduction, a preemphasis filter is applied to the speech signals to make it spectrally flattened. FIR filter is the most common filter used for this purpose, which is defined by Eq. (10.26).

$$H(z) = 1 - 0.95z^{-1} \tag{10.26}$$

which gives the preemphasized s(n) signal in the time domain as Eq. (10.27).

$$s(n) = \sum_{k=0}^{M-1} h(k)\hat{s}(n-k) \tag{10.27}$$

10.5.7.2 Endpoint detection

Endpoint detection is the most important part of the speech preprocessing. To find the exact location of the endpoints of a preemphasized speech signal is usually a challenging task in speech processing. An incorrect determination of the

10.5 Speech signal processing

end-points may reduce the performance of the system. Some techniques depend on measurements of the short-term power spectrum (P_s), short-term energy (E_s), and short-term zero crossing rate (Z_s). For a speech signal s(n) these parameters are defined in Eqs. (10.28) and (10.29).

$$E_s(m) = \sum_{n=m-N+1}^{m} s^2(n), \tag{10.28}$$

$$P_s(m) = \frac{1}{N} \sum_{n=m-N+1}^{m} s^2(n), \tag{10.29}$$

$$Z_s(m) = \frac{1}{N} \sum_{n=m-N+1}^{m} \frac{|sign[s(n)] - sign[s(n-1)]|}{2}, \tag{10.30}$$

where,

$$sign[s(n)] = \begin{cases} +1 \, for \, s(n) \geq 0 \\ -1 \, for \, s(n) < 0 \end{cases}$$

and N signifies the total count of samples present in one block of a signal. It is observed that the values of the power spectrum and short time energy increases with the speech present in the signal while the rate of short time zero crossing will decrease. For a better estimation about the beginning and the ending of the speech utterances, a trigger has to be generated based on background noise.

10.5.8 Frame blocking

The speech signal is generally defined as a nonstationary signal with quick variation in time and space. Therefore to observe its statistical behavior, the full preprocessed signal x(n) is required to be segmented into tiny fragments. Each tiny segmented fragment is called a frame, on which the speech sample is supposed to be approximately stationary with a stable statistical behavior. Speech seems to be stationary by choosing a frame size between 8 ms to 30 ms. The process of segmentation of a speech sample into tiny frames is called frame blocking.

The speech sample is blocked such that there will be some overlapping with the next frame in the blocked samples [24]. We have varied the percentage overlapping of the frame from zero overlapping to 75% overlapping. LPC features have been calculated first by no percentage overlapping of frames. In the case of overlapping, each blocked frame with N samples is overlapped with the neighboring frames by M samples. So the second blocked frame starts M samples after the previous frame and overlaps it by N-M samples; the third blocked frame starts 2 M samples after the first frame and overlaps it by N-2M samples, and so on. The overlapping percentage varies from model to model, subject to the application and how much speech data is required to be present in the speech signal. Frame overlapping helps to extract speech feature in a decidedly smoother fashion. The loss of signal details can be minimized with a higher amount of overlapping. This process is also useful

226 CHAPTER 10 Parameterization techniques for automatic speech

for better speech content extracted from a signal. There is no particular technique for determining the optimal overlapping percentage of frames.

10.5.9 Windowing

At the time of frame blocking, at the beginning and end of the block, there may be a possibility of signal discontinuity. At the next level, windowing is used to reduce the signal discontinuity at both ends of the frame. A special type of function known as window function is used in it. The window function minimizes the start and the end of the frame to zero and tapers down spectral distortion at both sides. Therefore when a speech signal s(n) is multiplied with a window w(n), it produces a set of speech samples x(n) which are weighted by time-interval and shape of the window. A simplest rectangular function is given by Eq. (10.31).

$$w(n) = \begin{cases} 1, 0 \leq n \leq N - 1 \\ 0, otherwise \end{cases} \tag{10.31}$$

Hamming window is another commonly used function whose nonzero part is a raised cosine pulse as described in Eq. (10.32).

$$w(n) = \begin{cases} 0.54 - 0.46cos\dfrac{2\pi n}{N - 1}, 0 \leq n \leq N - 1 \\ \\ 0, otherwise \end{cases} \tag{10.32}$$

where n is the number of samples in a given frame and N represents the total number of samples in the frame. A Hamming window function is represented in Fig. 10.9, in which the x-axis represents the sample number, and the y-axis indicates the amplitude of the samples. In this study, a window length of 10−50 ms is used for the speech analysis with 10 ms. The Hamming window is applied to 0% to 75% of overlapping speech segments.

10.6 Features for speech recognition

Speech signal has various features associated with it. These features ensure that the system accurately understands the word being spoken by the users. According to the study [25], speech features can be divided into three categories, as shown in Fig. 10.10.

10.6.1 Types of speech features

10.6.1.1 Spectral speech features

These features are dependent on frequency domain. These can be extracted using various techniques; the popular ones include linear predictive coding (LPC) and Mel frequency cepstral coefficient (MFCC).

10.6 Features for speech recognition

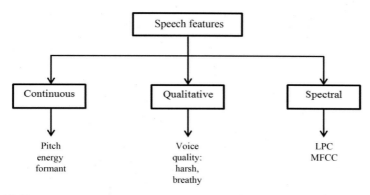

FIGURE 10.10
Categories of speech features.

1. LPC is a technique used to represent the spectral envelope of speech signals. The linear predictors are extracted using specific steps. It is a powerful technique that encodes speech into a numerical representation. It has further extended to LPCC and WLPCC.
2. MFCC is one of the most dominant features in speech classification. MFCC has a close analogy to the human ear and is based on social perception sensitivity; that is, it extracts those features that the human ear extracts for hearing purposes.

10.6.1.2 Continuous speech features

It includes speech features like pitch, formant, energy etc. These types of features provide a lot of relevant information in recognition of speech. These are time-dependent. Some of the features in this category are:

1. Pitch: it is also called fundamental frequency, which is the result of vibrations of the vocal cords. Pitch represents an essential characteristic of origin of input excitation [26]. The fundamental frequency in males ranges from 85 Hz to 155 Hz and in females, it ranges from 165 Hz to 255 Hz.
2. Formant: the speech produced by vibration of vocal cords manifests itself in an envelope, which is called the spectral envelope. The formant is the maximum value of this envelope.
3. Energy: it is the magnitude of loudness. It is also termed as volume or usually amplitude of the speech signal.

10.6.1.3 Qualitative speech features

The features in this category depict the quality of the speech produced. The voice utterances can be harsh, rough, and breathy. It affects all the other features extracted.

228 CHAPTER 10 Parameterization techniques for automatic speech

10.7 Speech parameterization

Feature extraction is the process of measuring speech-language content and acoustic variation. It is a representation of acoustic data. When the input signal is too large and redundant to be processed for an algorithm then dimensionality reduction is usually applied to it. Acoustic features are generally based on the short-term Fourier transform of an input speech signal. In this section, we will discuss the features, namely, linear predictive coefficients (LPC), linear-predictive cepstral coefficients (LPCC), weighted linear predictive cepstral coefficients (WLPCC), Mel-frequency cepstral coefficients (MFCC), and power spectrum density (PSD). Different parameter settings were used to achieve better results. However, as we are concerned to classify the speech of children with intellectual disabilities and typically developed children using supervised learning, features like LPC, LPCC, WLPCC, and MFCC are directly useful to apply as inputs to the model. For better classification, the best significant features are to be selected using feature selection algorithms and reduced representations sets, and are more reliable as an input vector. In the next section, we will discuss the extraction of the most significant feature vectors from the speech samples by varying various parameters: frame length, overlapping percentage, first order pre-emphasis filter and order in LPC, frame length, and number of filterbanks in MFCC, and downsampling frequency in PSD. Further, we will discuss the vector quantization method which converts a speech feature vector into a discrete numeric series.

10.7.1 Feature extraction

For extraction of useful features of a speech sample, many mathematics-based methods have been used and developed by researchers and scholars. In this study, linear predicative coding (LPC) is applied as a feature extraction technique whose mathematical function uses an all-pole model. LPC is a precise and straightforward method to implement and extract acoustics features analytically. This all-pole model has been proved to be the best approximation to the human vocal tract spectral envelope [27−29].

Given a speech signal of time n in Eq. (10.10), s(n) can be studied as the linear combination of the previous p samples as Eq. (10.33).

$$s(n) = b_0 u(n) + a_1 s(n-1) + a_2 s(n-2) + \ldots + a_p s(n-p) \tag{10.33}$$

It further equals Eq. (10.34).

$$s(n) = b_0 u(n) + \sum_{i=1}^{p} a_i s(n-i) \tag{10.34}$$

where u(n) stands for normalized excitation signal, b_0 is the gain of the same excitation signal, and $a_1, a_2, and a_n$ are the weights for the past sound signals.

10.7 Speech parameterization   229

On the speech processing frame, all these coefficients are considered as constants. Alternatively, s(n) represents the z-domain explained in Eq. (10.35).

$$S(z) = b_0 U(z) + \sum_{i=1}^{p} a_i S(z) z^{-1} \tag{10.35}$$

It gives a transfer function as:

$$H(z) = \frac{S(z)}{U(z)} = \frac{b_0}{1 - \sum_{i=1}^{p} a_i z^{-1}} \tag{10.36}$$

Given a sequence of data s(n), LPC predictive coefficients a_k are analyzed for minimizing the difference between the actual and the predicted output, i.e., the predictive error, shown in Eq. (10.37).

$$e(n) = s(n) - s^n = s(n) - \sum_{k=1}^{p} a_k s(n-k) \tag{10.37}$$

averaged on a windowed frame. However, the all-pole LPC model is considered to be a better estimation of the vocal tract spectral envelope, but these estimations are not appropriate for many real-time applications. When LPC coefficients are treated as features, it produces additional difficulties as they are not orthonormal. Therefore the LPC coefficients are converted into different spectral representations, namely, reflection coefficients, root pairs, and cepstrum coefficients [30,31]. In this research, we have used linear predictive cepstrum coefficients and weighted linear predictive cepstrum coefficients with linear LPC-based variables applied in many machine learning-based speech recognition systems. Further, these features are numerically convenient, efficient, and orthogonal.

In this section, three speech parameterization techniques are examined, namely, LPC, MFCC, and power spectrum density (PSD) on different parameters to extract the spectral features and apply them to classify the speech samples of children with intellectual disability and typically developed children. Linear predictive analysis is further categorized in LPC, LPCC, and WLPCC.

10.7.2 Linear predicative coding (LPC)

Linear prediction technique estimates the output of a linear system on the basis of its current input x_n and past output $s_{n-1}, s_{n-2}, \ldots, s_{n-p}$. Mathematically, it is shown in Eq. (10.38).

$$\hat{s}_n = \sum_{k=1}^{p} a_k s_{n-k} + \sum_{k=0}^{N} b_k x_{n-k} \tag{10.38}$$

where $\hat{s_n}$ describes the estimate of s_n. The major task is to compute the values of the coefficients a_k and b_k in such a way that the $\hat{s_n}$ approximates the actual output s_n, as precisely as possible.

230 **CHAPTER 10** Parameterization techniques for automatic speech

Depending upon the above equation, different types of models have been used, like the autoregressive model (AR), moving average model (MA), and autoregressive moving average model (ARMA). In the first model, the output $\hat{s_n}$ is estimated by using the past output samples and current input. In the second method, the prediction solely depends upon the input which gives $a_k = 0$ corresponding to an FIR filter. In the third method, the prediction is based on the general recursive filter [32,33].

10.7.2.1 Autocorrelation method

According to Eq. (10.39), the sum of the coefficients is finite over n due to the finiteness of the signal, but it is beneficial to think that there are very few samples of nonzero values as the frame is infinitely lengthy.

$$e_n = s_n - \hat{s_n} = s_n - \sum_{k=1}^{p} a_k s_{n-k} \tag{10.39}$$

Consequently, a windowed signal s_n is used where very few finite samples are nonzero which is described as

$$s_n = \begin{cases} somesampledsignal, 0 \le n \le N-1 \\ 0 \, n < 0 \, and \, n \ge N \end{cases}$$

The total mean square error should be minimized by Eq. (10.40).

$$E = \sum_{n} e_n^2 = \sum_{n} (s_n - \hat{s_n})^2 \tag{10.40}$$

Considering the above two equations and the windowed signal s_n, demonstrated in general for all time durations between $-\infty < n < \infty$, with $a_0 = 1$, we get,

$$E = \sum_{n=-\infty}^{\infty} \left(S_N - \sum_{k=1}^{p} a_k s_{n-k} \right)^2 = \sum_{n=-\infty}^{\infty} \left(\sum_{k=0}^{p} a_k s_{n-k} \right)^2 \tag{10.41}$$

By applying the partial derivative of E with respect to every a_i ($1 \le i \le p$) to zero, the total squared error can be reduced.

$$\frac{\partial E}{\partial a_i} = 2 \sum_{n=-\infty}^{\infty} \left(\sum_{k=0}^{p} a_k s_{n-k} \frac{\partial}{\partial a_i} a_k s_{n-k} \right) = 2 \sum_{k=0}^{p} a_k \sum_{n=-\infty}^{\infty} s_{n-k} s_{n-i} = 0$$

which can also be written as Eq. (10.42)

$$\sum_{k=0}^{p} a_k \gamma_{k,i} = 0 \tag{10.42}$$

where

$$\gamma_{k,i} = \sum_{n=-\infty}^{\infty} s_{n-k} s_{n-i} = \sum_{n=-\infty}^{\infty} s_{n+i-k} s_n = \sum_{n=-\infty}^{\infty} s_n s_{n-(k-i)} \tag{10.43}$$

With all these equations, we can observe that the term $\gamma_{k,i}$ depends on (k-i); that is why it can be stated as one variable autocorrelation function $\gamma_{k-i} = \gamma_{k,i}$. With the property of symmetry of γ_i, i.e., $\gamma_i = \gamma_{-i}$ and $a_0 = 0$, the minimized condition of total squared error is given by Eq. (10.44).

$$\sum_{k=1}^{p} a_k \gamma_{k-i} = -\gamma_i. \tag{10.44}$$

Here γ_i is the autocorrelation function of s_n with i = 1, 2, ..., p. The normal equation is defined by Eq. (10.44). By putting Eq. (10.44) into Eq. (10.41), we get the minimum total square error as Eq. (10.45).

$$E_p = \sum_{k=0}^{p} a_k \gamma_k \tag{10.45}$$

It is observed that the autocorrelation function γ_i and γ_{k-i} depends on additive limits. With the help of autocorrelation function, we can predict the AR parameters by solving a group of linear equations. The above equations are represented in the form of a matrix as shown in Eq. (10.46).

$$\begin{pmatrix} \gamma_0 \gamma_1 \gamma_2 \cdots \gamma_{p-1} \\ \gamma_1 \gamma_0 \gamma_1 \cdots \gamma_{p-2} \\ \gamma_2 \gamma_1 \gamma_0 \cdots \gamma_{p-3} \\ \cdot \\ \gamma_{p-1} \gamma_{p-2} \gamma_{p-3} \cdots \gamma_0 \end{pmatrix} \begin{pmatrix} a_1 \\ a_2 \\ a_3 \\ \cdot \\ a_p \end{pmatrix} = - \begin{pmatrix} \gamma_1 \\ \gamma_2 \\ \gamma_3 \\ \cdot \\ \gamma_p \end{pmatrix} \tag{10.46}$$

These p linear equations to be resolved are best analyzed in the form of a matrix as R x = y, where R stands for a p × p matrix and x is a column vector of LPC coefficients. To solve the LPC vector, it requires inversion of the R matrix and multiplication of the Y vector with the resultant p × p matrix with the resultant inversion matrix.

The calculation of the input and the output sequences produces a model with a number of poles or formants. The analysis then approximates the values of a discrete-time signal as a function of the linear combination of past samples. The spectral envelope is represented in a compressed form, using the information of the linear predictive model. Three acoustic features LPC, LPCC, and WLPCC are extracted from the speech samples as explained in Fig. 10.11.

10.7.3 Linear predictive cepstral coefficients (LPCC)

The speech sample can be observed as the convolution of the vocal tract signal with the excitation waveform. The spectral deconvolution technique is used to separate the envelope and the excitation components by transforming the multiplication of two spectra into the summation of two signals. The input sequence x(n) excites the linear time-invariant system and leads to an output sequence y(n).

FIGURE 10.11

Feature extraction using linear predictive analysis.

If y(n) is the output sequence of a linear time-invariant system being excited by an input sequence x(n), defined in Eq. (10.47).

$$Y(z) = X(z)H(z) \tag{10.47}$$

The logarithm gives Eq. (10.48).

$$C_y(z) = lnY(z) = lnX(z) + lnH(z) = C_x(z) + C_h(z) \tag{10.48}$$

It means that the cepstrum of output sequence y(n) is the summation of the cepstrum values of x(n) and h(n) and is shown in Eq. (10.49).

$$c_y(n) = c_x(n) + c_h(n) \tag{10.49}$$

which specifies that the convolution of two input sequence in the time domain results in the sum of the cepstral sequence in the cepstral domain. In practical, $c_h(n)$ eliminates after a few milliseconds and $c_x(n)$ is nonzero for voiced utterances for $n = 0, \pm N, \pm 2N, \pm 3N, \ldots$. The two functions are freely disjointed.

In LPC approach, the transfer function is defined as a p all-pole filter, shown in Eq. (10.50).

$$H(z) = \frac{b_0}{1 - \sum_{i=1}^{p} a_i z^{-1}} \tag{10.50}$$

A recursion relation can be established between the prediction coefficients a_k andcepstralcoeficnetsc_k by an all-pole model for H(z). By taking the derivative of Eq. (10.51).

$$c_n = \frac{1}{2\pi i} \int_C \ln H(z) z^{n-1} dz \tag{10.51}$$

where C stands for the closed contour about the origin, we can calculate the recursion as explained in Eq. (10.52).

$$c_1 = a_1$$

$$c_n = a_n + \sum_{k=1}^{n-1} \frac{k}{n} c_k a_{n-k}, 1 \leq n \leq p$$

$$c_n = \sum_{k=1}^{n-1} \frac{k}{n} c_k a_{n-k}, n > p \qquad (10.52)$$

where $a_1\ldots\ldots a_p$ are the LPC coefficients of order p and c_n stands for the corresponding first p values of the cepstrum. The visual representation of LPCC features of a child with intellectual disability and typically developed child is shown in Fig. 10.12A and B.

10.7.4 Weighted linear predictive cepstral coefficients (WLPCC)

Weighted linear predictive cepstral coefficients (WLPCC) can be determined by multiplying LPCC to the weighted formula, as explained in Eq. (10.53).

$$w_m = \left[1 + \frac{Q}{2}\sin\left(\frac{m\pi}{Q}\right)\right] 1 \leq m \leq Q \qquad (10.53)$$

In the cepstral region, a weighted function can be represented as a bandpass filter to deemphasize c_m around m = 1 to Q. The sensitivity of higher order cepstral coefficients onto the overall spectral scope at m = 1 and lower order cepstral coefficients to noise at m = Q are reduced by the filter. $\widehat{c_m}$ is calculated by Eq. (10.54),

$$\widehat{c_m} = c_m * w_m \, 1 \leq m \leq Q \qquad (10.54)$$

In previous studies, WLPCC has been applied as a technique to extract speech features in other applications with different neurodevelopment disorders [34].

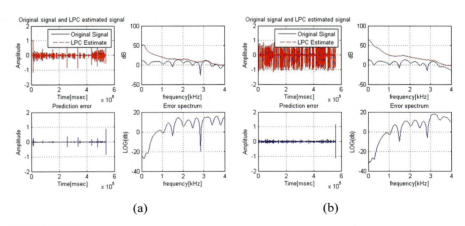

FIGURE 10.12

Visual representation of linear predictive cepstral coefficients (LPCC) of: (A) child with intellectual disability; or (B) a typically developed chld.

234 CHAPTER 10 Parameterization techniques for automatic speech

Table 10.3 Parameters setting for experiments.

Experiments	Frame size (ms)	α Value	Overlapping %	Order
Impact of frame size	10–50	0.97	0	12
Impact of α	20	0.91–0.99	33.33	12
Impact of % of overlapping	20	0.99	0–75	12
Impact of order	20	0.99	33.33	8–20

This method is extremely beneficial as it significantly decreases the feature space dimensions. On applying FFT or DFTs, the total number of output coefficients depends on the order of the FFT, whereas LPCC is based on the order of the pole p of the all-pole model so that the LPCC has a small picture of the spectral envelope. Biologically, these LPCC features show the difference between the physical structures of the person's vocal tract. The stages involved in the LPC-based feature extraction process are shown in Fig. 10.11. The LPCC values depend upon the number of variables, primarily, frame size (N), frameshift (M), and order (p) of the LPC analysis. Although to measure the effect on LPC, LPCC, and WLPCC, these variables can be varied over a wide range of values. This study has discussed the effect on the classification accuracy using different values of frame size, percentage of frame overlapping, α value of preemphasis first-order filter, and the order p. The parameters values are varied as mentioned in Table 10.3. The sampling rate Fs = 8000 Hz is used for the analysis. From each of the samples, the best 128 cepstral coefficients are extracted. This complete feature extraction process is coded in MATLAB 7.0.

The classification results were discussed with four combinations of speech parameters: frame size, order p, percentage of frame overlapping, and α value of first-order preemphasis filter of attributes of linear prediction. Four experiments were established, based on these experimental setups, and depicted in Table 10.3. In the first experiment, the effect of frame size was analyzed by fixing α to 0.97, no frame overlapping, and order value 12. In the second experiment, the effect of α was determined by setting the frame size to 20 ms (best-found value from the previous experiment), and the remaining parameters were at their previously used values. In the third experiment, the percentage of frame overlapping was analyzed by fixing the best values for frame size and α. The best value for the order was used. For the last experiment, the optimized value of the order p was determined by setting the other three parameters to their best values governed from previous experiments.

10.7.5 Mel-frequency cepstral coefficients

Mel-frequency cepstral coefficients are designed to represent the structure of a human auditory system. MFCC is a robust feature and commonly used in speech signal processing as it is inclined to the nonlinear perception of frequencies. Implementation of MFCC with multiple speakers with multiple languages is quite easy. MFCC obtained from the triangular-shaped filter banks is arranged in a frequency domain.

As explained earlier, human voice production is dependent on vocal filters. The speech signals vary in different person as it is affected by several parameters. i.e., language, socioeconomic background, and speech pathologies. It is influenced by the structure of the tongue, teeth, lips, and jaws; these are the natural filters of the speech production process. MFCC is a speech feature which can give the exact representation of the human auditory system. The only difference is that the MFCC uses the frequency domain representation and the speech signal uttered by the speaker is in a time domain. So to resolve this issue, MFCC takes the input speech signal in the time domain, converts it into the frequency domain for processing and, finally to represent the output, transforms the signal again into the time domain.

For all the speech recognition work presented in this chapter, we have used MFCC speech features in combination with LPC and spectral analysis. However, this section concentrates only on MFCC feature extraction along with its first and second derivatives. This section explains the MFCC feature extraction process with the significance of each step involved. All the codes for the feature extraction are carried out in MATLAB. Fig. 10.13 shows the steps of MFCC features extraction.

10.7.5.1 Preemphasis

The first step to compute the Mel-coefficients is preemphasis. The input signal may have some unbalanced high frequencies. So, preemphasis is applied to balance those high frequencies in the speech signal as there is a steep whirl in the high frequency parts of the voiced sound spectrum. In human speech recognition, the spectrum produced by voiced sounds has the glottis slope of around -12dB. Due to articulators like lips, tongue, jaw, the spectrum grows with a boost in the energy of about $+6$ dB. When the speech is recorded from a certain distance then there will be a descending slope of -6dB. Here, preemphasis removes the effect of glottis in the vocal tract [35].

FIGURE 10.13

Steps for extraction of MFCC feature.

In machine speech recognition, the speech signal is made to pass through a filter which balances or emphasizes the high frequency parts. The filter used for this purpose is given in Eq. (10.55).

$$Z(x) = Z(x) - \alpha * Z(x-1) \qquad (10.55)$$

The term $Z(x)$ is the output signal after the preemphasis process and α is the preemphasis coefficient ranging from 0.9 to 1.0. In this research, the value of α was taken as 0.97. The output signal from a preemphasis filter is shown in Fig. 10.14.

10.7.5.2 Framing

The second step comprises the breaking of the speech signal into small segments and then performing the further steps on each frame. The speech signal fluctuates with time, so it is better to perform the analysis on a small segment of the signal. This process is called framing. The signal (in seconds) is broken down into frames of milliseconds. The duration of each frame is 20 to 40 ms, if the duration of the speech signal is smaller or larger then the results may not be consistent. Some of the crucial terms used in the process are:

1. Frame size: it corresponds to how long a particular frame is, i.e., the frame duration.
2. Sampling rate or sampling frequency: it is the number of samples contained in each second. It always depends on the recording device.
3. Frame length: total number of samples contained in a frame.

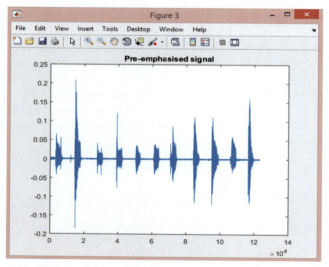

FIGURE 10.14

Preemphasized signal.

4. Frame shift: it provides the overlapping among the frames to avoid discontinuous edges.

Each parameter above can be computed using any other two parameters. For example, the frame length can be calculated using the sampling rate and frame duration.

$$Frame\ Length = Frame\ Duration * Sample\ Rate$$

In this work, the frame duration of 0.25 second or 25 ms and the sampling frequency of 8 kHz was used, so the number of samples in each frame was 200. The frame shift was 10 ms. The frame shift of the 10 ms had 80 samples. The first frame with 200 samples was started at sample 0 to sample 200; the second frame was started at sample 80 and ended at sample 280. Similarly, the third frame started at sample 160 until sampling 360 and so on.

10.7.5.3 Windowing

In the third step, the small signal called frame is then passed through a window. This window further processes the signal. This window is applied to maintain the continuity among all the frames from first to the last. A rectangular window is used to do this process and it is called the Hamming window. MATLAB has a built-in function named "hamming" to carry out the process of windowing. This function is represented in Eq. (10.56).

$$H(x) = \begin{cases} 0.54 - 0.46 * cos\left(\dfrac{2\pi x}{X-1}\right), if\ 0 \leq x \leq X - 1 \\ 0, otherwise \end{cases} \tag{10.56}$$

$H(x)$ is the windowed frame, where x is the current frame and X is the total number of frames in the signal. The function mentioned above is multiplied with each frame of the signal, that is, $Z(x)*H(x)$. Fig. 10.15 displays the generalized window formed after applying the above function on a signal.

10.7.5.4 Fast fourier transform

The fourth step is very important stage in the MFCC feature extraction. The fast Fourier transform function is applied to the windowed function to transform each frame of the signal from time domain to frequency domain. This process is shown in Eq. (10.57).

$$Z(k) = \sum_{n=1}^{N} Z(x)*h(x)e^{-2\pi kx/X}, if\ 1 \leq k \leq K \tag{10.57}$$

where $Z(k)$ is the signal in the frequency domain with length k.

Apart from this, this process makes MFCC features similar to way the the human voice is perceived by the ear. The working of the MFCC feature is similar to the structure of the human ear as well. The human ear has an organ named the

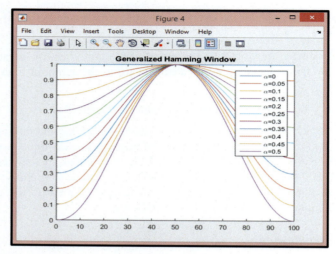

FIGURE 10.15

Hamming window.

"cochlea" that measures the presence of all the frequencies in the speech produced by the speaker. It contains a liquid substance where the vibration occurs at different locations depending upon the presence of articulated frequencies. These locations help the nerves in firing signals to the brain for the processing of speech recognition.

In the computer-aided speech recognition process, a similar thing is required in order to find out the presence of different articulated frequencies. This is called the estimation of the power of each frame or the "periodogram estimate," elaborated in Eq. (10.58).

$$P_i = \frac{1}{X}|Z(k)|^2 \qquad (10.58)$$

10.7.5.5 Mel-filter banking

After the conversion of the signal to the frequency-domain, this step is responsible for the computation of Mel-filter bank. The usefulness of the MFCC features entirely depends upon the Mel-filter banks. The reason being that the periodogram estimated in the above step contains enough information related to the speech signal. The "cochlea" is inefficient in identifying the two close frequencies, and this issue becomes more effective when the frequency starts increasing. For this purpose, the speech signal is passed through the filter banks that gets narrower or wider according to the frequencies. These filter banks are Mel-filters which are triangular band-pass filters [36]. The frequency (in Hz) is converted to Mel-scale using the Eq. (10.59) [37]:

$$M(f) = 1125\ln(1 + f/700) \tag{10.59}$$

After calculating the frequencies present in the signal, the Mel-frequency is converted back to the normal frequency by using the Eq. (10.60).

$$M^{-1}(m) = 700\left\{exp\left(\frac{m}{1125}\right) - 1\right\} \tag{10.60}$$

To ensure the closeness of the MFCC functions to the working of the human ear, we take the log value of the normal frequencies calculated in the previous step. In this study, in total 13 filter banks were used. The selected MFCC parameters for best outcome were as follows: frame length = 25 ms, frameshift = 20 ms, filter banks = 20, cepstral coefficients = 12, and frequency limit = 300−37 kHz. MFCC's 0th cepstral coefficient has been calculated as the average energies of all the filter banks [38]. Thirteen MFCC features were used to classify the speech of children with ID from TD children.

10.7.5.6 Discrete cosine transform

The purpose of this last step is to convert the signal back to the time-domain and also to decorrelate the filter banks for better classification [39]. The function to calculate the discrete cosine transform is mentioned in Eq. (10.61).

$$D_m = \sum_{k=1}^{M} \cos\left[(k - 0.5) * \pi/M\right] * E_k, form = 1 to C \tag{10.61}$$

where C is the total number of MFCC coefficients to be extracted and M is the number of Mel-filters used. The visual representation of MFCC features of a child with intellectual disability and typically developed child are shown in Fig. 10.16A and B.

10.7.6 Delta coefficients

The human voice has a lot of dynamic information associated to it. MFCC coefficients derived in the previous section are used as an input to calculate the first-order derivative, also called the delta coefficient. These coefficients are nothing but adjoining the trajectories of MFCC. They increase the recognition rate and accuracy and can be represented by Eq. (10.62).

$$d_t = \frac{\sum_{p=1}^{P} p(mfcc_{t+P} - mfcc_{t-P})}{2\sum_{p=1}^{P} p^2} \tag{10.62}$$

where dt is delta coefficients derived, and the standard value of P is 2.

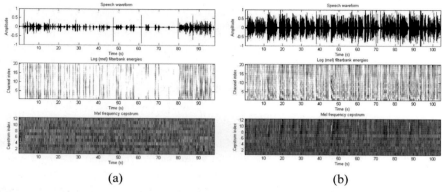

FIGURE 10.16

Visual Representation of Mel-frequency cepstral coefficients (MFCC) of: (A) child with intellectual disability, and (B) a typically developed chld.

10.7.7 Delta−delta coefficients

To further enhance the accuracy of the recognition system, the second derivative of MFCC is calculated using Eq. (10.61). The MFCC coefficients are replaced by the delta coefficients obtained using MFCC coefficients.

10.7.8 Power spectrum density

In speech signal processing, spectral density estimation aims to define the power spectral density of a continuous sequence of the time signal. Generally, the power spectral density functions characterize the frequency data of the signal [40]. The purpose of determining the power spectral density is to find the periodicities in the data, by detecting peaks of the frequencies of these periodicities. Spectrum analysis is a procedure of spiriting a complex speech signal into smaller parts. It can be applied on the entire signal, or the signal can be decomposed into small frames and then spectrum analysis can be performed on these frames.

The Fourier transform of a signal generates a frequency spectrum, consisting of all the data about the primary signal, but in some other form. It means that the primary speech signal can be readily synthesized by applying an inverse Fourier transform. To achieve a perfect reconstruction, the spectrum analyzer must hold two values: amplitude and phase of each frequency component. A two-dimensional vector represents these two variables as magnitude and phase in the form of a complex number. The resulting output is called a power spectrum.

In the frequency domain, Fourier transform is used to represents a function. The analysis in the frequency domain is simpler to perform as compared to a time domain. In real time, almost all software and electronic appliances that produce frequency spectra work on discrete Fourier transform (DFT), which runs on

10.7 Speech parameterization

samples of the speech signal. Fast Fourier transform (FFT) is used to implement the mathematical approximations of DFT [41,42].

The speech signal of each research participant (60 sec time interval), with silences between the words, were distributed into subsequent nonoverlapping time frames of 25 msec. Every frame had 256 data-points, which were transformed into the frequency domain with the help of 256 point FFT, DFT equation stated in Eq. (10.63). Then the power spectrum density is calculated using Eq. (10.64) by taking the square of the complex magnitude of the first 64 frequency channels which match to the frequency ranges from 0–2 kHz. The extraction of the spectral features is described in Fig. 10.17. The study focuses on the mentioned frequency range, as most of the acoustic-prosodic data are encoded here. A total of 64 spectral features were taken for building up the dataset.

For all time windows, the long-term average power spectrum (LTAPS) of the speech signals was calculated as the channel by channel average power spectra, as described in Eq. (10.65). The normalized LTAPS was calculated by dividing the power spectra in every frequency channel from the maximum power. This process compared the spectra of different objects [43,44]. For frequency channels, the average power was determined by using the following equation.

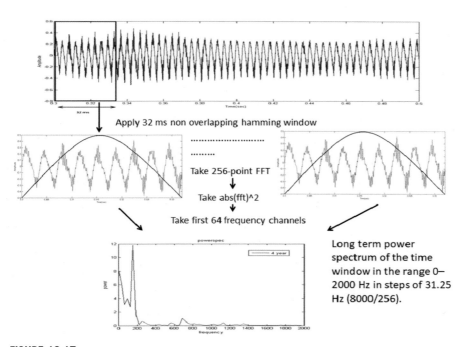

FIGURE 10.17

Extraction of spectral features.

242 CHAPTER 10 Parameterization techniques for automatic speech

$$X(i) = \sum_{n=0}^{N-1} x(n)e^{-j2\pi in/N} \tag{10.63}$$

where $i \in [0, N-1]$. Here $N = 256$ input sequence of complex number $x(n): = x_0, x_1,..x_{N-1}$. Transformed sequence of complex numbers $X(n): = X_0, X_1, ...X_{N-1}$

$$PSD_i = \{X(i)\}^2 \tag{10.64}$$

where PSD is defined as Power Spectrum Density at ith channel $i \in [1, 64]$

$$LTAPS_k = \left\{ \frac{\sum_{j=1}^{m} PSD_{j1}}{m}, \frac{\sum_{j=1}^{m} PSD_{j2}}{m}, ...\frac{\sum_{j=1}^{m} PSD_{j64}}{m} \right\} \tag{10.65}$$

where $LTAPS_k$ is defined as the kth speech sample and m is the number of time windows.

10.8 Speech recognition

The rapid technological advances in the past decades have increased the use of computers and smart devices, and according to a study [45], computers play an essential role in almost all the aspects of daily life. But not all the humans from rural areas are computer literate. So, a different approach was adopted, interaction using speech. Speech recognition is an integrative field of human—computer communication. It refers to the identification of the words being spoken by the user. A speech pattern recognition system is also used for the prediction and classification of neurodevelopmental disorders. A computer program can extract the features of a speech signal and predict the pathologies in the voice. This type of system can be used to identify speech—language disorder, and particular speech therapy will be suggested to the patient.

The first speech recognition system was introduced by Microsoft when they built an application that translated Chinese to the English language. Over the past few decades, many kinds of research have been going on in the field of speech recognition. Until 2001, the accuracy of the speech recognition systems measured was 80%, and up to 2010 there was no progress made. It was with the onset of Google Voice Search App that the development began to edge back up. Google tried to enhance the model accuracy and also added this feature to the Google Chrome browser [46].

Speech recognition has also paved the way for biometrics. Biometrics is a vast field that studies the human characteristics for their identification. Biometric has two classes associated with it, namely, physiological biometric identifiers that include identification using facial expressions, DNA, fingerprint, and another type called behavioral biometric identifiers, that include identification using voice and rhythm. The developments in the speech recognition technology have contributed to immense growth in the behavioral biometric identifiers. Almost all the smart

10.8 Speech recognition

devices used today have voice search, voice passwords, and speech-to-text converters.

The conventional speech recognition system has three primary stages associated with it, namely, preprocessing stage, feature extraction stage, and classification stage (it is often called pattern recognition stage). These three stages are explained in Fig. 10.18.

- **Preprocessing**: this process involves improvisation done to the speech signal to obtain a better voice signal. Preprocessing generally results in improving the quality of speech signal by using specific techniques and removing the silenced part or increasing the amplitude.
- **Feature extraction**: it involves extraction of apt information from the speech signal. In this step, the preprocessed signal is used as an input and output obtained is numerical representation of the speech signal. There are various features associated with a speech signal, and each one has different extraction techniques.
- **Classification**: this step is also called system modeling. It involves training and testing of the model according to the desired output. There exist various classification techniques involved in training and testing that will be discussed

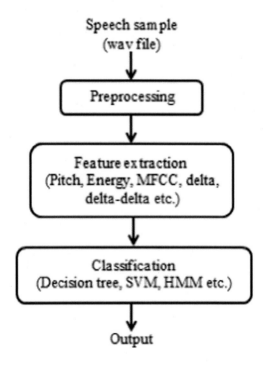

FIGURE 10.18

Conventional speech recognition system processes.

244 **CHAPTER 10** Parameterization techniques for automatic speech

in the sections ahead. Different classification algorithms are used depending on the performance of the model.

- **Basic terms:** speech recognition is a communication with a computer that accurately recognizes the words spoken by a person. Some commonly used terms in the speech recognition technology are described below:
  1. Phoneme: it refers to the smallest unit of sound used for formation of a word. A phoneme is enclosed within "/" at its both ends. For example, words "cat" and "kit" use phoneme /k/.
  2. Utterance: it refers to speaking a word (or words), sentence (or sentences) that is meaningful to the computer.
  3. Vocabulary: it represents the number of words on which a speech recognition system can work. Usually, these are further categorized as short, medium, large, and very large vocabulary, where the number of words is limited to tens in short, hundreds in medium, thousands in large, and ten thousand in extensive vocabulary.
  4. Training: it refers to making the system learn. It can be done by using classification modeling. It is possible because the speech recognition system can adapt to characteristics of new users.
  5. Accuracy: it depicts how well a speech recognition system works. It is improved by training the model.

10.8.1 Types of speech pattern recognition

The different types of speech recognition systems are:

1. Isolated word recognition: such a system can recognize only one utterance or one word at a time.
2. Connected word recognition: it is similar to isolated word recognition, but different utterances can run together with a short pause between them.
3. Continuous speech recognition: these systems allow the user to have communication with computers in the most natural way and are the most difficult ones to develop.
4. Voice identifiers: such systems are capable of identifying specific users through their voices. These are used in biometrics and where the security is concerned. Voice identifiers are also used to classify two-class problems like gender classification and classification of speech pathology.

10.9 Speech classification

Pattern recognition and the classification process are done using various algorithms. The speech recognition system can be embedded in multiple applications, which range from biometric to GPC systems. Classification involves categorization which can be statistical or predictive. The process begins with a dataset and some predefined classes. The dataset is used as a training set for the model which is then applied to the unseen dataset. For example, classifying the people by their eligibility for a loan will have two classes: either the person is eligible for getting a loan or not.

Similarly, the thesis focuses on the classification of children with intellectual disability from their peer controls using speech, which aims to classify the voice into two classes: intellectually disabled and typically developed. There are various classification algorithms available and the most common used for classification of speech signals are SVM, ANN, GMM, and HMM. The recent developments have shown enhancement toward the use of these algorithms by using combinations of various classification algorithms with different parameters values, for example, the combination of SVM and ANN is applied. Some of the algorithms are described as follows:

10.9.1 Artificial neural network (ANN)

This classifier studies the human brain to make or predict the decisions. The neural network indicates the set of inputs or outputs units which are connected and each group has a weight associated with it. The learning in the neural networks is accomplished by adjustment of these weights to predict the correct class of the input. Due to these connections among the units, the neural network learning is known as connectionist learning. The neural network is highly tolerant of the noisy data and is the best fit for prediction where there is little knowledge available about relationship among attributes [47,48]. Neural network technique also supports parallelism. There are different kinds of neural network algorithms available. Deep neural network (DNN), multi-layer perceptron (MLP), convolutional neural network (CNN), and recurrent neural network (RNN) are further categories of neural network.

10.9.2 Support vector machine (SVM)

It takes the set of input data and predicts the class to which each input belongs, by representing the dataset as points in space on the graph whose dimensions depend upon a number of classes. SVM finds a hyperplane using training dataset and some margin that is defined by the datasets. SVM is a highly flexible method as it can be used to distinguish between linearly separable and nonlinearly separable datasets (where there is no clear separation with a hyperplane between the classes). SVM can be used to identify or categorize the dataset among two or more than two categories. A hyperplane is the middle of the margin or the gap that accurately classifies the data. The datasets on the two margins are called support vectors. The margin distance has to be as maximal as possible—the different types of kernels that SVM supports are dot, radial, polynomial, neural, gaussian_-combination, multiquadric, ANOVA, and epachnenikov. These kernels have different operations and functions associated with them [49].

10.9.3 Linear discriminant analysis (LDA)

Linear discriminant analysis plays an integral part in feature selection and classification. LDA gives decent outcomes in many research fields such as speech processing, image processing, and gesture recognition. In the present study, linear discriminant analysis is applied to distinguish between the speech features of ID and TD into two classes. LDA is used to find a linear transformation that broadens the class separability margins in lower dimensional space [50].

10.9.4 Random forest

Random forests are ensemble learning classifiers used for classification tasks by building a forest of decision trees in a training task, resulting in the class that is the node of the classification or prediction of the single tree. An extended version of the algorithm was introduced by Leo and Adele which combines bagging with a random selection of attributes that builds a collection of decision trees with controlled variance. Random forest has been widely used for data classification in many applications like object recognition [51]. The forest here refers to an ensemble of decision trees that are trained using the bagging approach.

10.10 Summary and discussion

In this chapter, we perform the speech parameterization with linear predictive analysis (LPC, LPCC, and WLPCC), Mel-frequency cepstral coefficients (MFCC) and power spectral density feature. Also, the feature selection algorithm was applied and selected the most appropriate feature from the complete dataset. The best features are selected to decrease the computational complexity of the classifiers without affecting the classification performance. We performed an LPC analysis, which simulates the voiced speech signal as the output of an all-pole filter in response to the standard order of excitation pulses. LPC analysis is based on the categorization of all-pole filters. It is one of the vital tools in speech pattern analysis and recognition for low bit rate speech coding. The LPC has a compact representation of the spectral magnitude with simple computation. It is used to calculate vocal tract area function, fundamental frequency, and formant frequencies. LPC analysis of order varying from $8-14$ is performed which permits us to calculate LPC coefficients.

Further, the cepstral analysis was performed to convert the LPC coefficients into cepstral coefficients (LPCC). From each speech sample frame, 128 LPCC features have been extracted. The results suggested that the LPCC values are more significant than LPC. Further, the LPCC values are weighted by a window function tapered at both ends to estimate the weighted LPCC. Parameters such as window size, α, overlapping percentage, and order are varied over a wide range and the best values are extracted for each parameter.

Mel-frequency cepstral coefficients are commonly used features in speech pattern recognition, speaker identification, and processing. The shape of the human auditory arrangement establishes itself in the envelope of short-time power spectrum and MFCC exactly defines it. MFCC parameterization gives $1-13$ MFCC filter coefficients. Here, the speech samples can be confirmed as the convolution of the source excitation and vocal tract filter. The deconvolution can be attained by cepstral analysis of two components. Thirteen MFCC features were extracted to differentiate between the speech sample of children with intellectual disability and typically developing children. For better results, various features of MFCC such as frame length and value of α were varied with a nonoverlapping window. The best result was considered with frame length $= 25$ ms, frame shift $20 =$ ms, $\alpha = 0.97$, number of filter bank channels $= 20$, cepstral coefficients $= 12$, cepstral sine lifter parameter $= 22$, and frequency limit $= 300-3700$.

In speech signal processing, spectral density estimation aims to define the power spectral density features of a continuous sequence of the time signal. Generally, the power spectral density functions characterize the frequency data of the signal. The purpose of determining the power spectral density is to find the periodicities in the data, by detecting peaks of the frequencies of these periodicities. Spectrum analysis is a procedure of spiriting a complex speech signal into smaller parts. It can be applied on the entire signal, or the signal can be decomposed into small frames and then spectrum analysis can be performed on these frames. The power spectrum density is calculated with downsampling of 8 kHz and 16 kHz, respectively. We want to inform that a significant difference has been found between the speech signals of children with intellectual disability and typically developed children. Three experiments have been conducted to classify the speech signals of children with intellectual disability and typically developed children.

The early diagnosis of intellectual disabilities has never been easy, specially for developing nations. In this research, models based on machine learning supervised algorithms are used to distinguish between speech of children with intellectual disabilities and normally developing \kids. Three different types of speech features are applied to build the database: MFCC, LPCC, and power spectral features.

The review of this research holds the findings of comparable studies on speech and language development. Mervis et al. in 2007 examined the speech of 167 children with ID and recognized that 71.3% of these individuals were affected by acoustic and language diseases. However, this study did not examine the auditory nature of those neurodevelopmental disorders. The study adds to the results by studying the character of neurodevelopmental disorders through vocal markers. The research also delivers evidence of vitality of acoustic markers in prediction by training the classification algorithms which will categorize between speech of typically developing children and youngsters with intellectual disabilities. Speech impairments in children with intellectual disability can arise because of deficits in listening or speaking or both. Also, it is a challenge to indicate that the acoustic skills are overdue in ID as these deficits are also observed in adults with ID [52,53].

This research delivers a concept that fine motor skills, such as speech, have proven to be an early indicator that can be used to identify ID children. However, the scope of research is always there. The study can be taken to another level by incorporating more speech features and type. Video and gesture recognition can also be used to increase the reliability of the classifiers. The research should be carried out on a large speech dataset with participants from diverse socioeconomic backgrounds and races. The latest machine learning algorithms like convolutional neural network and recurrent neural network can be used for more significant outcomes.

Reference

[1] American Psychiatric Association, Diagnostic and statistical manual of mental disorders (DSM-5®), American Psychiatric Pub, 2013.

[2] P.K. Maulik, M.N. Mascarenhas, C.D. Mathers, T. Dua, S. Saxena, Prevalence of intellectual disability: a meta-analysis of population-based studies, Res. Dev. Disabilities 32 (2) (2011) 419–436.

[3] E. Keller, The cortical representation of motor processes of speech, Mot. Sens. Process. Lang. (1987) 125–162.

[4] B. Dodd, Differential diagnosis and treatment of children with speech disorder, John Wiley & Sons, 2013.

[5] J. Piaget, Language and thought of the child: selected works, vol 5, Routledge, 2005.

[6] L. Wing, Language, social, and cognitive impairments in autism and severe mental retardation, J. Autism Dev. Disord. 11 (1) (1981) 31–44.

[7] Organisation mondiale de la santé, World Health Organization, WHO Staff, WHO, The ICD-10 classification of mental and behavioural disorders: clinical descriptions and diagnostic guidelines, vol. 1, World Health Organization, 1992.

[8] A.M. Liberman, F.S. Cooper, D.P. Shankweiler, M. Studdert-Kennedy, Perception of the speech code, Psychological Rev. 74 (6) (1967) 431.

[9] O.C. Ai, M. Hariharan, S. Yaacob, L.S. Chee, Classification of speech dysfluencies with MFCC and LPCC features, Expert. Syst. Appl. 39 (2) (2012) 2157–2165.

[10] M.A. Anusuya, S.K. Katti, Front end analysis of speech recognition: a review, Int. J. Speech Technol. 14 (2) (2011) 99–145.

[11] H. Beigi, Fundamentals of speaker recognition, Springer Science & Business Media, 2011.

[12] Paul Cuff. ELE 201: Information signals - course notes, 2015. <http://www.princeton.edu/~cuff/ele201/kulkarni.html>. 59.

[13] Macquarie University. Speech waveforms. <http://clas.mq.edu.au/speech/acoustics/waveforms/speech_waveforms.html>.

[14] WebExhibits. Newton and the color spectrum. <http://www.webexhibits.org/colorart/bh.html>.

[15] B.P. Bogert, The quefrency alanysis of time series for echoes; Cepstrum, pseudo-autocovariance, cross-cepstrum and saphe cracking, Time Ser. Anal. (1963) 209–243.

[16] Oppenheim, A.V. (1965). Superposition in a class of nonlinear systems.

[17] O. Cappé, Elimination of the musical noise phenomenon with the Ephraim and Malah noise suppressor, IEEE Trans. Speech Audio Process. 2 (2) (1994) 345−349.

[18] O. Cappé, J. Laroche, Evaluation of short-time spectral attenuation techniques for the restoration of musical recordings, IEEE Trans. Speech Audio Process. 3 (1) (1995) 84−93.

[19] Yang, J. (1993). Frequency domain noise suppression approaches in mobile telephone systems. In *Acoustics, speech, and signal processing, 1993. ICASSP-93., 1993 IEEE International Conference on* (vol. 2, pp. 363−366). IEEE.

[20] J.H. Hansen, M.A. Clements, Constrained iterative speech enhancement with application to speech recognition, IEEE Trans. Signal. Process. 39 (4) (1991) 795−805.

[21] Lockwood, P., Boudy, J., & Blanchet, M. (1992). Non-linear spectral subtraction (NSS) and hidden Markov models for robust speech recognition in car noise environments. In *Acoustics, Speech, and Signal Processing, 1992. ICASSP-92., 1992 IEEE International Conference on* (vol. 1, pp. 265−268). IEEE.

[22] J.C. Junqua, The Lombard reflex and its role on human listeners and automatic speech recognizers, J. Acoustical Soc. Am. 93 (1) (1993) 510−524.

[23] M.G. Rahim, B.H. Juang, W. Chou, E. Buhrke, Signal conditioning techniques for robust speech recognition, IEEE Signal. Process. Lett. 3 (4) (1996) 107−109.

[24] A.M. Goberman, M.P. Robb, Acoustic examination of preterm and full-term infant cries: The long-time average spectrum, J. Speech, Language, Hearing Res. 42 (4) (1999) 850−861.

[25] M. El Ayadi, M.S. Kamel, F. Karray, Survey on speech emotion recognition: features, classification schemes, and databases, Pattern Recognit. 44 (3) (2011) 572−587.

[26] Q. Liu, M. Yao, H. Xu, F. Wang, Research on different feature parameters in speaker recognition, J. Signal. Inf. Process. 4 (02) (2013) 106.

[27] L.R. Rabiner, B.H. Juang, Fundamentals of speech recognition, vol. 14, PTR Prentice Hall, Englewood Cliffs, 1993.

[28] R.C. Snell, F. Milinazzo, Formant location from LPC analysis data, IEEE Trans. Speech Audio Process. 1 (2) (1993) 129−134.

[29] Campbell, J., Tremain, T. (1986). Voiced/unvoiced classification of speech with applications to the US government LPC-10E algorithm. In *Acoustics, Speech, and Signal Processing, IEEE International Conference on ICASSP'86.* (vol. 11, pp. 473−476). IEEE.

[30] Jakobson, R.F., Fant, G., Halle, M. (1963). Preliminaries to speech analysis: the distinctive features and their correlates.

[31] B. Gold, N. Morgan, D. Ellis, Speech and audio signal processing: processing and perception of speech and music, John Wiley & Sons, 2011.

[32] J.D. Markel, A.H. Gray, Linear prediction of speech, Springer Verlag, New York, 1976.

[33] J.A. Cadzow, ARMA modeling of time series, IEEE Trans. Pattern Anal. Mach. Intell. (2)(1982) 124−128.

[34] M. Hariharan, L.S. Chee, S. Yaacob, Analysis of infant cry through weighted linear prediction cepstral coefficients and probabilistic neural network, J. Med. Syst. 36 (3) (2012) 1309−1315.

[35] K.S. Rao, S.G. Koolagudi, Robust emotion recognition using spectral and prosodic features, Springer Science & Business Media, 2013.

[36] P. Dhanalakshmi, S. Palanivel, V. Ramalingam, Classification of audio signals using SVM and RBFNN, Expert. Syst. Appl. 36 (3) (2009) 6069–6075.

[37] S. Jothilakshmi, V. Ramalingam, S. Palanivel, Unsupervised speaker segmentation with residual phase and MFCC features, Expert. Syst. Appl. 36 (6) (2009) 9799–9804.

[38] J.W. Picone, Signal modeling techniques in speech recognition, Proc. IEEE 81 (9) (1993) 1215–1247.

[39] Erokyar, H. (2014). Age and gender recognition for speech applications based on support vector machines.

[40] Stoica, P., Moses, R.L. (2005). Spectral analysis of signals.

[41] P. Welch, The use of fast fourier transform for the estimation of power spectra: a method based on time averaging over short, modified periodograms, IEEE Trans. Audio Electroacoustics 15 (2) (1967) 70–73.

[42] D.B. Percival, A.T. Walden, Spectral analysis for physical applications, Cambridge University Press, 1993.

[43] I. Lehiste, N.J. Lass, Suprasegmental features of speech, Contemporary issues Exp. phonetics 225 (1976) 239.

[44] X. Xiao, E.S. Chng, H. Li, Normalization of the speech modulation spectra for robust speech recognition, IEEE Trans. Audio, Speech Lang. Process. 16 (8) (2008) 1662–1674.

[45] Hofmann, N. (2014). Speech recognition for human computer interaction. In *ETH Zurich, hofmannn@ student. ethz. ch, Ubiquitous computing seminar FS.*

[46] S. Swamy, K.V. Ramakrishnan, An efficient speech recognition system, Computer Sci. Eng. 3 (4) (2013) 21.

[47] C. Neocleous, C. Schizas, Artificial neural network learning: a comparative review, Hellenic conference on artificial intelligence, Springer, Berlin, Heidelberg, 2002, pp. 300–313.

[48] J. Schmidhuber, Deep learning in neural networks: an overview, Neural Netw. 61 (2015) 85–117.

[49] S.B. Kotsiantis, I. Zaharakis, P. Pintelas, Supervised machine learning: a review of classification techniques, Emerg. Artif. Intell. Appl. Comput. Eng. 160 (2007) 3–24.

[50] C.H. Park, H. Park, A comparison of generalized linear discriminant analysis algorithms, Pattern Recognit. 41 (3) (2008) 1083–1097.

[51] Räsänen, O., Pohjalainen, J. (2013). Random subset feature selection in automatic recognition of developmental disorders, affective states, and level of conflict from speech. In *INTERSPEECH* (pp. 210–214).

[52] G. Aggarwal, L. Singh, Classification of intellectual disability using LPC, LPCC, and WLPCC parameterization techniques, Int. J. Comput. Appl. (2018) 1–10.

[53] G. Aggarwal, L. Singh, Evaluation of supervised learning algorithms based on speech features as predictors to the diagnosis of mild to moderate intellectual disability, 3D Res. 9 (4) (2018) 55.

CHAPTER

Impact of big data in healthcare system—a quick look into electronic health record systems

11

Vijayalakshmi Saravanan[1], Ishpreet Aneja[2], Hong Yang[2], Anju S. Pillai[3] and Akansha Singh[4]

[1]*Faculty in Department of Software Engineering, Rochester Institute of Technology, Rochester, NY, United States*
[2]*Department of Data Science, Rochester Institute of Technology, Rochester, NY, United States*
[3]*Department of Electrical and Electronics Engineering, Amrita School of Engineering, Amrita Vishwa Vidyapeetham, Coimbatore, India*
[4]*Department of CSE, ASET, Amity University Uttar Pradesh, Noida, India*

11.1 A leap into the healthcare domain

Health care data originally was kept in the form of hard copies but with recent trends, health systems are racing to digitize these large amounts of patient health data. The information includes medical diagnoses, prescriptions, allergy data, demographics, clinical narratives, and the clinical laboratory results [1]. With the shift to digitization, we have observed improvements in healthcare system with data being widely available with reduced examinations, fewer ambiguities caused by illegible handwriting, fewer missing or lost patient records, etc. Patient care has immensely improved with healthcare systems becoming more organized.

Electronic health records are not the only data that are being recorded. With advancements in healthcare technology, multiple instruments have transformed into digital forms. Like electronic health records which were originally static, X-ray films, and scripts all went digital. Small electronics such as heart rate monitors, EKG, oxygen sensors, and large electronics such as MRI machines are continuously recording data. Keeping all these streams of data organized is a serious challenge, especially with each data being different types. The challenge of any healthcare system is, how to keep everything, including MRI images, sensory data, and patient health records, organized and interlinked. The historical relational databases that currently exist will struggle with maintaining structured and unstructured data.

Data analytics revolve around the 4 Vs of data characteristics. These are volume, velocity, variety, and veracity, which will be the main challenges health

---

Machine Learning and the Internet of Medical Things in Healthcare. DOI: https://doi.org/10.1016/B978-0-12-821229-5.00009-4
Copyright © 2021 Elsevier Inc. All rights reserved.

care must overcome in order to become dominant in data organization and analytics [2]. Health care is continuously creating and accumulating data at an exponential rate. Healthcare systems will have to attempt to store this large volume of data and allow access at any time with a high rate of velocity. This brings us to the second data characteristic, velocity, which will be needed to analyze data in real-time. In cases of real-time monitoring of ICU blood pressure, operating room anesthesia measurements, heart monitors, etc. it can be the difference between life and death. Variety is a big challenge in health care since the data streams come from multiple structured and unstructured sources. Health systems will have to understand and manage these streams of data, such as MRI, X-ray images, sensory data, along with genomic data. Healthcare applications will have to be efficient in combining and converting a variety of forms of data, including conversion from structured to unstructured data or other creative data storage solutions. Veracity is the last challenge that health care will need to overcome. Is the data reliable and captured correctly? Is the data secure from outside intrusion/attack? Methodology and solutions will have to be constructed to tackle all these challenges.

There are additional challenges that are in health care as well. These include the high cost to overhaul the entire IT infrastructure that exists currently in the healthcare system. Not many healthcare systems have the money or the resources to dedicate to this kind of technological overhaul. Not only is the cost high but there needs to be resources and a number of inputs required to pull off a successful project. The healthcare system will have to organize doctors, scientists, software engineers, and data scientists all come together and bring about a full technological reconstruction, which is a huge challenge by itself.

The end goal of health care should be to successfully maintain multiple streams of data and grow to include genomic data, in order to provide individualized healthcare treatment for patients and the highest quality of care possible. With the digitization and combining data in order to use big data, healthcare organizations ranging from single physician offices, multiple-provider groups to large hospital networks can receive significant benefits. Data would be more readily available; resulting in quick disease detection, and guaranteeing appropriate and spontaneous treatments. Population health would vastly improve, billing fraud would be detected quicker, and health systems would increase profits and reduce waste and redundant cost. Overall disease patterns and outbreaks would be heavily analyzed to improve surveillance and speed response. Health systems can use this data to deliver faster targeted vaccinations, such as the yearly influenza strain. Data collected would turn into actionable information to identify needs, provide services, predict and prevent crises for the benefit of the population. Enhanced data and analytics would benefit patients who are the greatest consumers of health resources and could be provided with factual and accurate information to make informed decisions, proactively manage their own health, and adopt and track healthier behaviors, as well as to identify treatments in real-time.

11.2 The real facts of health record collection

Due to the large volume of data that health systems will have to parse and explore, they will have to rely on distributed frameworks to divide and analyze the large amount of data. Open source projects like Hadoop and Apache Spark can be run on the cloud and provide a variety of analytics that healthcare systems can use. These open source projects empower health systems to analyze large data sets that were previously impossible to do. The data also have to be aggregated in a location where we can analyze them effectively and efficiently. Databases that are part of NoSQL technology packages, such as couchDB and MongoDB, offer solutions to pool the raw data where they can be loaded and analyzed by Hadoop, which has the ability to process large volumes of data with various structure or no structure at all [3]. There are numerous vendors such as Amazon AWS, Cloudera, and MapR technologies that distribute Hadoop platforms. The alternative Apache Spark is supported by SQL and is an in-memory processing of data which makes it much faster than Hadoop but comes at a cost for large datasets. Apache Spark can be used for real-time data solutions that healthcare systems require.

Healthcare data also consist of signals from data such as electrocardiograms, images, and video which are stored in a patient's electronic health records. The combination of images, videos, and structured texts stored in a patient's health records can be harnessed using artificial intelligence (AI). AI programs can draw actionable insights from the wealth of structured and unstructured data to make informed decisions and diagnose diseases. Healthcare professionals can take a look at the abnormalities provided by these machine learning approaches.

Image analytics is a whole category that healthcare can dive into with the abundance of CT, MRI, X-ray, molecular imaging, ultrasound, PET, EEG, and mammograms offering an abundance of imaging data that consist of large sizes. Radiologists and doctors do an excellent job of manually analyzing and finding abnormalities. However, there are many rare diseases and undiscovered diseases out there that can make diagnoses a challenge. To help in these situations machine learning can be used to recognize disease patterns from the large data set amassed over the years. There are also prebuilt classification libraries that have already analyzed millions of labeled imaged data that can assist doctors and healthcare professionals to diagnose patients correctly without the need for healthcare workers to manually train the algorithms themselves.

Healthcare systems can also collect data from large technological companies such as Google and Apple. With the latest trend of wearable devices, patients want to collect as much data as possible regarding their health. These wearable devices can have additional attachments for diabetes patients or cancer patients to record additional data to promote their health and wellness. Google and Apple provide developer kits that can allow health systems to tap into these data stores and keep doctors connected with their patients. The combination of wearable data

254 **CHAPTER 11** Impact of big data in healthcare system—a quick look

and existing patient health records can provide additional insight and personalized healthcare solutions for a patient [4]. Doctors can better manage patient conditions with wearables and track conditions and offer a more one-on-one treatment.

Health systems can also purchase a prebuilt commercial platform that is ready to use and user-friendly compared to an open source custom build solution healthcare system. A powerful and well-known platform is IBM Watson which is a commercial software that can share and analyze data among hospitals, providers, and researchers [5]. This commercial platform has the ability to extract maximum information from minimal input. Healthcare systems can easily set up these commercial out-of-the-box solutions and have continuous support from the company to troubleshoot any issues. These platforms are also validated and regulated for commercial release. Healthcare systems will have to pay the price if they choose to go this route but will have the benefits of a successful working analytics product [6].

11.3 A proposal for the future

The purpose of healthcare systems to adopt big data solution is to make their patients' lives easier and healthcare more efficient. To do so, the health system will have the option to purchase a prebuilt commercial solution where data can be stored and analyzed for actionable insights for the benefits of the healthcare system and the patient or the healthcare system starts from scratch and creates a custom-built solution that can be built and improved upon for years to come.

The custom-built solution will have to solve the volume, velocity, variety, and veracity problems the health system is facing. Starting with volume, the healthcare system will have to look into a storage solution. Many organizations prefer to keep data storage on premises to keep control over the data and ensure the up time. However on-site server networks can be expensive to scale and maintain over a period of time, especially as the amount of data grows at an exponential rate. Decreasing costs and increasing reliability makes cloud-based storage an appealing option for data solutions. Healthcare companies ideally should have a hybrid solution which is the most flexible and workable approach for various data access and storage needs.

Once we have the storage systems up and running, we have the beginning foundational layer of data analytics laid down. The next steps would be compiling the variety of data sources that we need to analyze. Healthcare data can come in many flavors and there are multiple options to store this data, e.g., a NoSQL database, such as mongoDB or couchDB, that can store enormous amounts of data collected from a healthcare system. Healthcare systems can also have the opportunity to deploy APIs to transmit data, especially when data is housed in different locations dependent on where testing is completed. Once data are compiled, we can use multiple choices of big data platforms and tools to begin performing big

11.4 Discussions and concluding comments 255

data analytics. The crowd favorites are Hadoop and MapReduce to parse through large amounts of data and deliver queries and reports. Big data algorithms can also be applied in real-time for data preprocessing purposes.

Healthcare systems should be able to develop a data pipeline to store large data collected from imaging, such as MRI and CT scans, and store them in a format suitable for big data platforms such as Spark. This data conversion process saves time for downstream data cleaning and allows health systems to analyze large amounts of imaging data for quicker analysis. Other machine learning algorithms can be applied at this stage to parse through unstructured data to find patterns and anomalies in patient health data. These algorithms can be trained and deployed to help prevent disastrous disease outcomes for patients and to assist doctors to deliver better care [7].

Using a custom-built solution for health care and deploying multiple open source products come at a security risk. Healthcare systems will have to invest against malicious attacks intended to steal patient data. Security systems will have to be fortified and follow the HIPAA security rules to store, transmit, and authenticate data and to control access for specific individuals [3]. Common sense security measures like using up-to-date antivirus software, firewalls, encrypting sensitive data, and multifactor authentication will have to be deployed. Proper architectural solutions will be recommended, especially patient sensitive data, such as social security, credit card information from billing, and patient test data, will have to be encrypted and data access restricted to a limited number of employees. Machine learning algorithms can also be deployed to detect any network anomalies and alert hospital systems if the network is facing any type of cybersecurity attack with countermeasures deployed to prevent any access to patient data [8].

11.4 Discussions and concluding comments on health record collection

Healthcare systems' biggest challenge is conquering the variety of data and developing a pipeline to analyze multiple datatypes simultaneously to gain insight into its patient health data, billing, clinical testing etc. The opportunities are endless for healthcare systems and the first hurdle most of them deal with is where to begin. There is a large cost of investment required to analyze large quantities of data. Most healthcare systems' management would not invest in such areas since there is very little liquid cash flow available. There is a fear of failed investments which would bring loss to the organization which would be catastrophic and a larger challenge financially to recover from.

Resources are another hurdle that healthcare systems will face. Full-scale data analytics for healthcare will need input from multiple sources including doctors, scientists, software engineers, and data scientists, which are all costly to employ.

256 **CHAPTER 11** Impact of big data in healthcare system—a quick look

These resources will have to spend time in planning, building, validating, and ensuring that the system is delivering the correct information. The output of these analytics will need to be verified to ensure correct information is being relayed to the patients. The wrong treatment or options would be a costly error for patients and would be a blowback for healthcare analytic projects as a whole.

Big data is built around open source projects such as Hadoop and Apache Spark. These open source projects are much cheaper than the box solutions that IBM Watson provides but these solutions come at a risk. There is not much technical support and a great deal of programming skills is required to successfully implement these tools. There is also a lack of security for open source frameworks which is a risk healthcare system must assess before they adopt such technologies. Healthcare systems are recommended to reinforce their network security to prevent a data breach [9].

The goal of big data in healthcare is to provide better treatment and a personalized diagnosis for patients. Healthcare systems can reduce cost in frequent testing and reduce improper diagnosis along with patients getting the best possible treatment much faster. The investment for big data will reduce costs in the long run and keep healthcare operations running more efficiently than ever. Both patients and healthcare systems will reap the rewards of big data.

11.5 Background of electronic health record systems

Developments in information technology have allowed organizations in every industry to transition from paper records to digital records. To better understand the electronic health record, we can look at a similar concept widely used in manufacturing. Many companies have integrated their business records and processes into an enterprise resource planning system (ERP). The process of digitization brings many benefits, with one report showing that 95% of companies improved their key business processes by implementing an ERP system [10]. Companies engage in this process of digitizing the records and the business processes of their company for many reasons, including to improve business performance and to make employee jobs easier [10]. In short, digitization of records and processes is the next step forward in improving the effectiveness of an organization. However, healthcare organizations face a different set of challenges than manufacturers and retailers.

11.5.1 The definition of an electronic health record (EHR)

The most succinct definition of an EHR can be taken from the Office of the National Coordinator for Health Information Technology, "An electronic health record (EHR) is a digital version of a patient's paper chart" [11]. However, this hides the various other requirements necessary for an electronic health record to function properly. EHR's must be real time, secured, and instantly accessible to authorized users from multiple healthcare organizations [12]. They must also "Contain a patient's medical history, diagnoses, medications, treatment plans, immunization dates, allergies, radiology images, and laboratory and test results" [11]. This is no small feat. While an ERP system performs well for standardized processes within an organization, the EHR must handle patient specific idiosyncrasies across multiple organizations.

Regardless of the challenges, the United States Government recognized the need to digitize health records to better serve the public. That is why they found the Office of the National Coordinator for Health Information Technology in 2004 to coordinate and oversee the implementation of effective IT in health care [12]. Central to all of this, is the electronic health record.

11.5.2 A short history of electronic health records

The usage of electronic health records in healthcare organizations has dramatically increased since 2004. With adoption rates greater than 80%, any patient in the United States has a very high chance of interacting with physicians using an EHR system [13]. There is good reason for this, as many studies show improvements in patient outcomes for healthcare organizations that invest into EHR systems. EHR systems have been linked to an increased influenza vaccination rate in inpatient settings from 1% to 50% [14], a reduced risk of nonintercepted serious medication errors by 50% [15], and a reduced history of malpractice claims by 40% [16]. These benefits are significant and many experts believe that ubiquitous use of EHR systems will provide even more benefits [17]. Overall, EHR systems have improved the quality of care and organizational efficiencies of healthcare organizations across the United States [18].

However, the wave of EHR adoption created many problems for physicians and healthcare organizations. There are challenges in the collection and input of EHR data that negatively impact practicing physicians in various ways. These challenges need to be resolved to fully realize the benefits of EHR. This case study will review three studies on the challenges facing EHR systems.

11.6 Review of challenges and study methodologies

In this section we present the review of three major challenges faced by EHR.

11.6.1 Analyzing EHR systems and burnout

A study published in the Journal of the American Medical Informatics Association identified a correlation between physician stress, burnout, and other factors [19]. Their methodology involved identifying which physicians worked in

258 CHAPTER 11 Impact of big data in healthcare system—a quick look

a high EHR environment and then surveying their stress, burnout, and other indicators. To identify which; physicians worked in a high EHR environment, the study surveyed clinic managers on the number of features available in their EHR system and the number of years of EHR use. EHR features include displaying lab results, writing prescriptions, exchanging data with other physicians, and many more. Generally, these are features in EHR systems that the physician is expected to use to carry out their job. This information on EHR features was used to classify clinics into three groups: high EHR, moderate EHR, and low EHR. Generally, high EHR clinics had more features and used EHR systems for longer, an average of 4 years compared to 1 year for low and moderate EHR.

Once the study determined which level of EHR the physicians were using, they surveyed the physicians on time pressure during patient visits, control over workplace issues, job satisfaction, job stress, burnout, and intent to leave the practice. Job satisfaction, burnout, and intent to leave were measured on a five-point scale. This type of methodology may reveal a correlation rather than causation. For example, high EHR clinics may attract younger tech savvy physicians that report higher stress than older physicians. The study did control for physician age, sex, specialty, work hours, and years using the EHR, but there may be other confounding variables.

However, using a survey to measure satisfaction, stress, and burnout is most likely the best available method for gathering data. This is because these factors are subjective and can only be directly measured by the individual going through that experience. Because the study surveyed 379 physicians and 92 clinics, the study's methodology should be sufficiently robust to apply to clinics and physicians in the United States.

The study identified that there was correlation of reduced job satisfaction by approximately 0.4 points in the high and moderate EHR clusters compared to the low EHR cluster. This finding has a very low P value of 0.01. The study also identified less significant increases in stress and burnout in the high and moderate EHR clusters compared to the low EHR cluster. There was no significant difference between clusters when it came to intent to leave. The researchers also provided insight on why moderate EHRs may be the most stressful and cause the most burnout. They believe that moderate EHRs systems are in a state of transition, where some processes are still using physical records. This state of transition involves process changes and inefficiencies in moving between paper and digital records. In contrast to the high EHR clinics, the moderate EHR clinics have been using EHR systems for fewer years on average (4 years vs. 1 year) and thus may need time to adjust.

11.6.2 Analyzing EHR systems and productivity

A study published in the Journal of the American Medical Informatics Association found that physicians believed that they lost valuable time due to the adoption of EHR systems [19]. This is concerning because the purpose of EHR systems is to improve care, not reduce efficiency and waste a physician's time. The study issued a survey of over 800 attending physicians and trainees. Approximately half of invitees responded in a valid manner, but there is a large proportion of trainees (25%). The survey tracks the subjective opinion of the respondents on the impact that EHR had on their productivity and efficiency. The study did not mention any controls for age, sex, experience, or any other factor.

This methodology may be problematic as physicians may have a negative bias against EHR systems. The wording of the survey may cause respondents to focus on how EHR systems caused loss of time, rather than gain in time. Furthermore, EHR gains in efficiency may be less salient or memorable because they do not cause problems. It is wholly possible that EHR systems cause a few large problems, but also provide many small benefits that add up. Respondents of the survey may only focus on the large problems, because the small benefits may be too many to count or easy to forget.

The study reports a high statistical likelihood that physicians believe that they lost almost 50 minutes per clinic day. This is a very large amount of time, but it is important to take the methodology of the study into consideration. This study does not conclude that EHR systems reduce efficiency. Instead, it convincingly concludes that physicians believe that EHR systems reduce efficiency. Because the study did not measure actual throughput values, such as the number of patients seen or time spent on each patient, we cannot conclude that EHR systems reduce real efficiency in a clinic.

However, the perceived loss of efficiency warrants further study. It is possible that the high proportion of trainees (25%) in this study influenced the overall result. Further studies should control for confounding factors such as age, sex, experience with EHR systems, experience as a physician, and other factors not listed here. It is also critical to study the real impact of EHR systems on productivity in addition to the perceived impact.

11.6.3 Analyzing EHR systems and data accuracy

Data accuracy is a major concern for EHR systems, but it is rare for software systems to lose data or return incorrect data. Oftentimes, inaccurate data is caused by incorrect use of the software system or incorrect expectations of what the system can handle. Both are detrimental to the overall effectiveness of any system. EHR systems suffer from data accuracy issues as well. A study published in the Journal of the American Medical Informatics Association identified high rates of error in some, but not all EHR systems and EHR tasks [5].

Unlike the previous two studies, this study measured physician interactions with EHR systems in a training setting. Because the study is not using a survey of physician opinions, it is less susceptible to individual bias and perception issues. A total of four sites were selected for the study. All sites used an EHR system from one of the two industry leaders in EHR. Six fictitious scenarios were

260 CHAPTER 11 Impact of big data in healthcare system—a quick look

designed and physicians ran through the scenarios with standardized instructions. No guidance was given during the scenarios. For each site, between 12 and 15 individuals were recruited for the study, with the study covering four sites and 55 individuals.

This methodology is intended to simulate a real-life situation and provide an opportunity for experimenters to measure a proxy of real-world performance. Unlike normal usage of EHR systems, this study provided closed off environments without potential distractions or stressors normally present in the workplace. Additionally, it is possible that the scenarios presented are hard to understand or unclear for the study's participants. However, a simulation of EHR system usage is a good way to estimate real-world EHR system usage without putting patients at risk.

The study reported high variation in error rates between tasks and sites. One important example is the prescription of drugs. One site reported a 0% error rate, while another reported an error rate of 30%. These errors include issuing the wrong dose, frequency, and/or other elements of a prescription. Because different sites often reported large differences in error rates for tasks, it is likely that the scenarios designed by the experimenters were not a confounding factor. Rather, if one site using an EHR system could achieve a 0% error rate, why should another site report a significantly higher error rate? Overall, the study identifies that EHR systems, even EHR systems from the same vendor, may perform very differently. These differences may be caused by factors such as a difference in customizations for each site, a difference in training for each site, a difference in the study participants at each site, and/or other factors not listed above.

The conclusion of the study is fairly convincing. In short, the study finds that EHR systems may be highly inaccurate in a simulated test setting. Further study is necessary to determine whether or not these inaccuracies are observed in real-life settings. It is also important to study whether or not these inaccuracies lead to harm for patients. Even though the EHR system may be inaccurate, physicians may be able to use their professional judgement to intervene and prevent harm.

11.7 Conclusion and discussion

Of the three studies reviewed, two studies used a survey of physicians and one used a simulation of EHR usage. While a simulated test of a system will provide more concrete evidence, such studies may be significantly more resource intensive. One important factor that all three studies misses is the direct measurement of an EHR system's real-world performance. It is still unclear whether or not EHR systems cause burnout, stress, errors, and inefficiency. Additionally, the only study to utilize a nonsurvey methodology did not include a comparison for non-EHR. It is possible that EHR systems report high rates of error and that non-EHR systems report even higher rates of error.

However, it should be acknowledged that studying the impact of EHR systems is very complex. For example, implementing an EHR system may provide an audit trail of physician errors, but does that mean that the EHR system is causing more errors? It is possible that the EHR system simply improves reporting of such errors. It is also difficult to study the real-world metrics of EHR systems without putting patients at risk. It may be unethical to split patients between a control group and an EHR group for treatment. Even so, it is critical for the health industry to understand the true impact of EHR systems. Overall, more research is necessary in the field of EHR systems.

Unfortunately, it is often not possible to fully understand a new technology before implementation. However, now that EHR systems are widely used in the United States and many other parts of the country, it is critical to research and understand their real impact. In this chapter, we studied how big data and its techniques are contributing toward the promising research domain of the healthcare system. We also emphasized the challenges in collecting and maintaining electronic healthcare records in real-time systems

References

[1] S. Dash, S.K. Shakyawar, M. Sharma, S. Kaushik, Big data in healthcare: management, analysis and future prospects, J. Big Data 6 (1) (2019). Available from: https://doi.org/10.1186/s40537-019-0217-0.

[2] S. Sarraf, M. Ostadhashem, Big data spark solution for functional magnetic resonance imaging.arXiv.org, 2016. <https://arxiv.org/abs/1603.07064>.

[3] B.K. Kalejahi, S. Meshgini, A. Yariyeva, D. Ndure, U. Maharramov, A. Farzamnia, Big data security issues and challenges in healthcare. arXiv.org, 2019. <https://arxiv.org/abs/1912.03848>.

[4] J. Padikkaparambil, C. Ncube, K.K. Singh, A. Singh, Internet of Things technologies for elderly health-care applications, Emergence of pharmaceutical industry growth with industrial IoT approach, Academic Press, 2020, pp. 217—243.

[5] R.M. Ratwani, E. Savage, A. Will, R. Arnold, S. Khairat, K. Miller, et al., J. Am. Med. Inform. Assoc 25 (2018) 1197—1201.

[6] C.J. Mcdonald, F.M. Callaghan, A. Weissman, R.M. Goodwin, M. Mundkur, T. Kuhn, JAMA Intern. Med. 174 (2014) 1860.

[7] N.R. Jyothi and G. Prakash, "A Deep Learning-Based Stacked Generalization Method to Design Smart Healthcare Solution", Lecture Notes in Electrical Engineering. 545 (2019) 211—222.

[8] W. Raghupathi, V. Raghupathi, Big data analytics in healthcare: promise and potential. *Health Inform. Sci. Syst.* 2 (1) (2014). <https://doi.org/10.1186/2047-2501-2-3>.

[9] E. Kolker, V. Özdemir, and E. Kolker, "How Healthcare Can Refocus on Its Super-Customers (Patients, n = 1) and Customers (Doctors and Nurses) by Leveraging Lessons from Amazon, Uber, and Watson.", OMICS. 20 (6) (2016) 329—333.

[10] 2018 ERP report, Panorama consulting solutions, (2019), <https://cdn2.hubspot.net/hubfs/2184246/201820ERP20Report.pdf>.

262   **CHAPTER 11** Impact of big data in healthcare system—a quick look

[11] What is an electronic health record (ehr)?, The office of the national coordinator for health information technology, (2019), <https://www.healthit.gov/faq/what-electronic-health-record-ehr>.

[12] Ehr adoption rates, The office of the national coordinator for health information technology, (2019), <https://www.healthit.gov/topic/about-onc>.

[13] Office-based physician electronic health record adoption, The office of the national coordinator for health information technology, (2019), <https://dashboard.healthit.gov/quickstats/pages/physician-ehr-adoption-trends.php>.

[14] P.R. Dexter, S. Perkins, J.M. Overhage, K. Maharry, R.B. Kohler, C.J. Mcdonald, N. Engl. J. Med. 345 (2001) 965−970.

[15] M. Singh, S. Sachan, A. Singh, K.K. Singh, Internet of Things in pharma industry: possibilities and challenges, Emergence of pharmaceutical industry growth with industrial IoT approach, Academic Press, 2020, pp. 195−216.

[16] A. Virapongse, D.W. Bates, P. Shi, C.A. Jenter, L.A. Volk, K. Kleinman, et al., Arch. Intern. Med. 168 (2008) 2362.

[17] A. Chapko, B. Lebreton, T. Finet, Enterprise interoperability II, 621−624.

[18] N. Menachemi Collum, Risk management and healthcare policy, 47 (2011).

[19] S. Babbott, L.B. Manwell, R. Brown, E. Montague, E. Williams, M. Schwartz, et al., J. Am. Med. Inform. Assoc. 21 (2014). Available from: 10.1136/amiajnl-2013-001875.

Index

Note: Page numbers followed by "*f*" and "*t*" refer to figures and tables, respectively.

A

AAL. *See* Ambient Assisted Living (AAL)
Accuracy, 28, 100
  of algorithm, 98
  of speech recognition, 244
Acoustic features, 228, 231
Acute diseases, 69
Adaptive filter, 52–53
  algorithms for ECG signal, 48, 58
  configuration, 53f
Adaptive voting algorithm, breast cancer using, 193
Adencarcinoma, 201
Administrative services, 149
Aggregation of data, 8–9
Agnihotra, 200
AI-based techniques, 89
AIME. *See* Artificial Intelligence in Medicine (AIME)
Algorithms, 4
  artificial intelligence, 72, 78f
  machine learning, 1–2, 24–25
All-pole LPC model, 228–229
Alphabet Incorporation, 18–19
ALS. *See* Amyotrophic lateral sclerosis (ALS)
Alzheimer's disease, 38, 94
Amazon machine learning, 15–16
Ambient assisted living (AAL), 136, 146
Amplitude, 217
Amyotrophic lateral sclerosis (ALS), 146
Analog signals, 216
Analog to digital (A/D) converter, 216
Animal intelligence, 72
ANN. *See* Artificial neural network (ANN)
Apache mahout, 16
Apache Spark, 253
Apple, 253–254
Architecture, of machine learning, 4–12
  data acquisition, 5, 6f
  data modeling, 9–12, 11f
  data processing, 5–9
  deployment, 12, 14f
  execution (model evaluation), 12, 13f
  phases of, 5f
ARMA. *See* Autoregressive moving average model (ARMA)
Arrhythmia, 60–61

Articulators, 211
Articulatory organs, 211
Artifact, types of, 52
Artificial intelligence (AI), 1, 156–157
  actions of, 75f
  algorithms, 72, 78f
  definition, 74–75
  disease. *See* Disease
  in healthcare, 73
    applications, 23–24
  history, 72–75
  in human services, 157–158
  medicine, 70–73. *See also* Medicine/drug
Artificial Intelligence in Medicine (AIME), 73
Artificial limb services, 146
Artificial neural network (ANN), 59–60, 83, 245
  layer in, 84f
AtomNet, 18–19
Augmented limb leads, 48
  electrodes, 50t
Autocorrelation method, 230–231
Autodiagnosing system, 121–122
Autoimmune disease (Ads), 67–69, 68f
Automated medication system, 145–146, 150
Automatic speech recognition system
  data collection, 214–215
    noise reduction, 215
    recording procedure, 214–215
  features for, 226–227, 227f
    continuous speech features, 227
    qualitative speech features, 227
    spectral speech features, 226–227
  intellectual disability, 209–210
  motivation, 210–211
  speech classification, 244–246
  speech parameterization, 228–242
  speech production, 211–213
  speech recognition, 242–244
  speech signal processing. *See* Speech signal processing
Automation complacency, 41
Autoregressive model (AR), 230
Autoregressive moving average model (ARMA), 230

B

Background analysis, 37
Back-propagation neural network (BPNN), 62

263

264 Index

Bar-coded pneumatic transport system, 149
Baseline wandering noise, 52
Bayes theorem, 103, 205
Beat classification method, 62
Bed management system, 148
Behavioral biometric identifiers, 242–243
Big data, 89–90, 113, 156
  analytics, 115–116, 116*f*
    in healthcare, 122–123
  healthcare and, 113–115, 114*f*
  and IoT applications, 158–159
Bill management system, 148
Biological nervous system, 82
Biomedical imaging, 156
Biomedical signal processing, 47
  electrocardiogram signal. *See also*
    Electrocardiogram (ECG) signal
    interference signals in, 52
    preprocessing of, 51–58
    reviews of, 48–50
  ML-based technique in, 47–48
Biometrics, 242–243
Biostatistical theory of disease, 67
Black-box decision-making, 39–40, 40*f*
Blockchain, 39
Blood pressure (BP), 102
BMI. *See* Body mass index (BMI)
Body mass index (BMI), 161
BPNN. *See* Back-propagation neural network
  (BPNN)
BrainScript language, 17
Breast cancer
  algorithm, 204–205
  case study in, 200–203
    discourse, 202
    history and assessment of patients, 201
    recommendations for diagnosis, 201–202
  dataflow, 196
  diagnosis, 105–106
    performance evaluation of, 106
  duty PDR and PER, 194–195
  illustration, 200*f*
  info structure, 195
  input stage, 195
  in IoHTML, 193–200
  output design, 195
  responsible developers overview, 195
  software development life cycle, 193–194,
    194*f*
  symptoms of, 198
  types, 198–200
  using adaptive voting algorithm, 193
BSN-Care, 182

C

Camcorders, 153
Cancer
  detection, 188, 189*f*, 201*f*, 203*f*
  in healthcare, 192–193
    methods, 192
    result, 192–193
  prediction, 188, 189*f*, 196, 197*f*
    in activity view, 196, 197*f*
    in case view, 196
    in class view, 196–197, 198*f*
    in state chart view, 198, 199*f*
Cardiotocography (CTG), 118, 125
Cardiovascular diseases, 91, 105, 160
Cardiovascular heart disease (CHD),
  161–162
Case-based data segregation, 5
Central processing units (CPUs), 15
Cepstral analysis, 222–223
Cepstral mean subtraction (CMS) method, 224
Cepstrum, 222
  of discrete signal, 222–223
CH. *See* Community Healthcare (CH)
CHD. *See* Cardiovascular heart disease (CHD)
Chest pain (CP), 102
Chest, precordial leads placement on, 50, 50*f*
CHI. *See* Children Health Information (CHI)
Children Health Information (CHI), 137
Chronic diseases, 31, 69
Classification, 93–94, 205*f*, 243–244
  algorithm, 58–59
  data mining and, 36–37
  feature extraction and, 58–62
  model, 206*f*
  neurodevelopmental disorders, 209, 211
  pattern recognition and, 244–245
  of speech sentences, 210
  technique, 2–4, 96–97
    reinforcement learning, 4
    supervised learning, 2–3
    unsupervised learning, 3–4
  using SVM, 99*f*
Cleft, 187
Cloud computing, 139, 141–143
Clustering algorithms, 4
CMS method. *See* Cepstral mean subtraction
  (CMS) method
CNN. *See* Convolutional neural network (CNN)
Cognitive sensomotoric, 210
Communication
  modes, 209
  speech disfluency in, 209
Community Healthcare (CH), 137

Index   265

Competitive advantage, 140
Complex algorithm techniques, 74
Computer-aided speech recognition process, 238
Conceptual diagnostic model, 107
Conditional Restricted Boltzmann Machine
    (CRBM) model, 94
Confusion matrix, 90, 97−98, 97t, 107,
    129−130
  for classification model, 130f
Connected word recognition, 244
Connectionist learning, 245
Continuous speech
  features, 227
  recognition, 244
Conventional algorithms, 2
Conventional healthcare system, 90−91
  challenges in, 91
Conventional speech recognition system,
    243−244, 243f
Convolutional neural network (CNN), 19, 80−82,
    80f, 90
Convolutional noise, 224
Coordination, in medical section, 139
Coronary artery disease (CAD), 161−163
  counteractive action, 163
  hypertension, 162
  insulin, 162
  treatment, 162
Corporate hospital, main problem area of,
    137−140
  coordination in medical section, 139
  cost leadership model in market, 140
  diagnostic services, 138
  inpatient services, 138
  location, 138
  medical record keeping, 139
  outpatient services, hassle on, 138
  support and utility services, 139
  transparency, 139−140
Correlation of diabetes, 168f, 170f
Cost leadership
  model in market, 140
  with quality of care, 148−149
Cost reserve funds, 159
CPUs. *See* Central processing units (CPUs)
CRBM model. *See* Conditional Restricted
    Boltzmann Machine (CRBM) model
Creating systems, 37
Crowdsourcing, 158
Crucial features, 129f
Crude health information, 158
Custom-built solution, 254
Cyber bullying, 37

D

Dartmouth conference, 72−73
Data, 39
  analytics, 127
  collection stage, 126
  life cycle of, 114, 114f, 126−127
  preprocessing stage, 126−131
  reduction and transformation stage, 126−127
  splitting, 127−131
Data acquisition, 5, 6f
Data analysis, 7, 251−252
Data cleaning, 8
Data communication, 122
Data consistency, 8
Dataflow, 196
Data formatting, 8
Data gathering, 118
Data mining, 16, 36−37
  techniques, 122, 187
Data modeling, 9−12, 11f
Data preprocessing, 7−8, 7f, 95
Data processing, 5−9, 7f, 10f
  analysis, 7
  arrangement, 6−7
  preprocessing, 7−8, 7f
  transformation, 8−9
Data sampling, 8
Data scientist, 8
Data segregation, 6
Dataset, 95, 100, 214, 244−245
  CSV files of, 125f
  research, 214t
  training and testing, 95−96
Data transformation, 8−9, 95, 118
Deafness, 160
Decision support system, 105
Decision tree (DT), 101−103, 102f
Decomposition of data, 8
Deep analysis, 188
Deep CNN. *See* Deep convolutional neural
    network (Deep CNN)
Deep convolutional neural network (Deep CNN), 60
DeepGene, 200
Deep learning system, 16−17, 90, 94, 104,
    106−107
Deep neural networks (DNN), 60, 92
Deep Q-network (DQN), 94
Deficiency diseases, 69, 69t
Degenerative diseases, 69, 69t
Delayed error normalized LMS (DENLMS)
    algorithm, 54
Delta coefficients, 239
Delta−delta coefficients, 240

DENLMS algorithm. *See* Delayed error normalized LMS (DENLMS) algorithm
Department of Information Technology (DIT), 119−120
Dependent variables, 83
Deployment, 12, 14*f*
Depression, 160
DFT. *See* Discrete Fourier transform (DFT)
Diabetes, 159−161
   effects of, 160−161
   and headache, 161
   prediction, 186, 186*f*
   with TTH, 168*f*, 170*f*
Diagnose heart disease, 185
Diagnosis disease, 89. *See also* Disease diagnosis system, machine-learning
   disease identification and, 157
   for heart disease, 186*f*
   outcomes of, 203
   recommendations for, 201−202
   and treatment, concept of, 70
   using machine learning, 37−41
      Alzheimer's disease, 38
      heart diseases, 38
      hepatitis, 38
      thyroid disease, 38
Diagnostic errors, 91
Diagnostic services, 138
Dietary services, 147
Digital health, medical imaging used in, 155−156
Digital signal, 216
Digital system, 216
Direct automated intervention, 144*f*
Discrete cosine transform, 239
Discrete data, 6
Discrete Fourier transform (DFT), 221, 240−241
Discrete Time Fourier Transform (DTFT), 221−222
Disease, 67−70
   autoimmune diseases, 68−69, 68*f*
   biostatistical theory of, 67
   classification of, 69
   concept of, 70*f*
   diagnosis and treatment, 70
Disease diagnosis system, machine-learning, 95
   algorithm, 98−104
      decision tree, 101−103
      K-nearest neighbors, 100−101, 101*f*
      naive Bayes algorithm, 103−104
      support vector machine, 98−100, 99*f*
      types, 93−95, 93*f*
   breast cancer, 105−106
   cardiovascular disease, 105

   confusion matrix, 97−98, 97*t*
   healthcare industry, 106−107
   neurological disorders, 104
   prediction, model for, 95−97, 96*f*
      classification technique, 96−97
      data preprocessing, 95
      performance metrics, 97
      training and testing data set, 95−96
   tools for, 91−92
      MATLAB, 93
      Python, 92, 92*f*
DIT. *See* Department of Information Technology (DIT)
Diverse intelligent systems, 95
DNN. *See* Deep neural networks (DNN)
Domain analysis, 32−37
Drugs
   discovery, 190
      and manufacturing of, 18−19
      process, 75−77
      and production, 157
Drug therapy, 70
DTFT. *See* Discrete Time Fourier Transform (DTFT)
Ductal carcinoma in situ (DCIS), 199

E

ECG signal. *See* Electrocardiogram (ECG) signal
Echocardiography, 156
E-commerce, 35
ECP. *See* Embedded context prediction (ECP)
EEG. *See* Electroencephalography (EEG)
Efficiency, 28
EGC. *See* Embedded gateway configuration (EGC)
E-healthcare systems, 33
EHRs. *See* Electronic health records (EHRs)
Einthoven's equilateral triangle, 48, 49*f*
Einthoven's law, 48
Electrocardiogram (ECG) signal, 47
   adaptive filters, 52−53
      algorithms for, 48
   augmented limb leads electrodes, 50*t*
   classifier-based statistical learning algorithm, 60
   feature extraction and classification, 58−62
      artificial neural network (ANN), 59−60
      fuzzy logic, 60−61
      hybrid approach for, 62
      wavelet transform, 61−62
   lead electrodes, 49*t*
   nonadaptive filters, 52−53
   preprocessing of, 51−58
      delayed error normalized LMS algorithm, 54
      least mean square algorithm, 54
      log least mean square algorithm, 57−58

normalized least mean square algorithm, 54, 56*f*

sign data least mean square algorithm, 55–56, 57*f*

processing, 50, 51*f*

reviews of, 48–50

transmission and data acquisition of, 51–52

Electrodes, 48, 49*f*, 49*t*

ECG augmented limb leads, 50*t*

Electroencephalography (EEG), 104

Electromagnetic radiation, 153

Electronic health records (EHRs), 91–92, 251, 256–257

and burnout, 257–258

challenges, 257–260

and data accuracy, 259–260

definition of, 256–257

impact, 261

and productivity, 258–259

short history of, 257

Electronic medical records, 139, 143*f*, 191*f*

cloud-based storage, 144*f*

Eloquent speech, 209

Embedded context prediction (ECP), 137

Embedded gateway configuration (EGC), 137

Emergency system, 149

EMG noise, 52

EMH. *See* Indirect emergency healthcare (EMH)

Endpoint detection, 224–225

Energy, 35

Enterprise resource planning system (ERP), 256

Environmental noise, 223–224

ERP. *See* Enterprise resource planning system (ERP)

Error signal, 53

ES. *See* Expert systems (ES)

Euclidean metric, 100–101

Execution stage, ML architecture, 12, 13*f*

Expert systems (ES), 72, 82–83

fuzzy, 83

Extended logistic regression model, 94

Eye harm, 160

F

False positive/false negative result, 41

Fast Fourier transform (FFT), 62, 237–238, 240–241

Faulty data, 28

FCM clustering. *See* Fuzzy c-means (FCM) clustering

Feature engineering, 9, 10*f*

Feature extraction, 95, 228–229, 243

block diagram for, 216*f*

and classification, 58–62

artificial neural network (ANN), 59–60

fuzzy logic, 60–61

hybrid approach for, 62

wavelet transform, 61–62

mel-frequency cepstral coefficients, 235, 235*f*

spectral, 241*f*

techniques, 210–211

using linear predictive analysis, 232*f*

Feature selection, 121

FFT. *See* Fast Fourier transform (FFT)

Financial transactions, 36

Flu, 70

Fluent speech, 209

F Measure, 98

Formant, 227

Fourier Series (FS), 220

Fourier techniques, 219–220, 220*t*

Fourier Transform (FT), 220–221, 240–241

Frame blocking, 225–226

Framing, 225, 236–237

Frequency analysis, 219–221

Frequency transformation, 219

F1 score, 98

Fully connected layers, 82

Functional margin, 99

Fundamental frequency, 227

Fuzzy-based inference system, 60

Fuzzy c-means (FCM) clustering, 60–61

Fuzzy expert systems, 83

Fuzzy logic (FL), 60–61

Fuzzy neural network, 62

Fuzzy systems, 82

G

Gaining insights, 118

Gaussian mixture models (GMM), 93

Gaussian Naive Bayes model, 103–104

Gaussian noise, 54

Generalization, 83

Geometric margin, 99

Gestational diabetes, 161

Glottis, 211

GMM. *See* Gaussian mixture models (GMM)

Goldwave Noise Reduction Technique, 215

Google, 19, 34, 117, 253–254

GPUs. *See* Graphic processing units (GPUs)

Graphic processing units (GPUs), 15

H

Hadoop, 254–255

Hadoop Distributed File System (HDFS), 126

Hamming window function, 223*f*, 226, 237, 238*f*

HCML, 193

268 Index

HDFS. *See* Hadoop Distributed File System
    (HDFS)
Headache, 161
Healthcare delivery system, 135
  corporate hospital. *See* Corporate hospital, main
    problem area of
  internet of things, 136–137
    cost leadership with quality of care, 148–149
    implementation of, 140–149
    and medical record, 142–143
    in patient delight, 148
    in supportive and utility services, 146–148
    and therapeutic facilities, 144–146
Healthcare industry
  machine learning in, 106–107
  need of IOT in, 116–117
Healthcare, in machine learning, 30–32, 36,
    135–136. *See also* Telemedicine
  analysis of domain, 32–37
  application of IOT in, 182–185
  artificial intelligence in, 23–24, 73
  background and related works, 32–33
  big data and, 113–115, 114*f*
    IoT applications, 158–159
  cancer in, 192–193
  community projects for, 34
  detection and prediction of cancer, 188, 189*f*
  existing applications, 33–37, 34*f*
    e-commerce, 35
    energy and utilities, 35
    finance, 36
    healthcare, 36
    manufacturing, 36
    retail, 34
    social media, 37
    transport, 36–37
  future perspective, 38–41
  integration of, 23, 33
  internet of things in, 136–137
  issues with, 40*f*
  machine intelligence in, 157–158
  nutshell, 24–25
  perspective of disease diagnosis using, 37–41
  redefining, 183–185
  role of big data analytics in, 122–123
  techniques and applications, 25–28
    desired features of, 28–29
    reinforcement learning, 27–28, 28*f*
    supervised learning, 25, 27*f*
    types of, 26*f*
    unsupervised learning, 26, 27*f*
  use in, 117, 185–192
    diabetes prediction, 186, 186*f*

    diagnose heart disease, 185
    discovery of drugs, 190
    liver disease prediction, 187, 187*f*
    study and clinical trial, 191–192
    surgery on robots, 187
    treatment tailored, 188–189
  working, 29–30
Healthcare systems, machine learning in, 17–19,
    182, 251–252, 254, 256
  applications, 18–19
    discovery and manufacturing of drugs,
     18–19
    identification and diagnosis of disease, 18
    medical images, 19
  domain, 251–252
  electronic health record systems, 256–257
  health record collection, 253–256
  proposal for, 254–255
Health informatics, 119
Health information, 171
Health record collection, 253–254
Health tracking, 159
Heart diseases, 38, 105
  diagnose for, 186*f*
Hepatitis, 38
Heterogeneous tumor, 105
Hidden Markov models (HMM), 93
HMM. *See* Hidden Markov models (HMM)
Homographic filtering, 222
Homomorphic filtering, 222
Hospital-related injuries, 146
HTN. *See* Hypertension (HTN)
Human intelligence, 72
Human–machine interaction, 210–211
Human speech production system, 212, 212*f*, 213*f*
Hybrid approach for ECG signal, 62
Hyperplane, 245
Hypertension (HTN), 162
Hypertrophy, 161
Hypoglycemia, 161

I

ICA. *See* Independent component analysis (ICA)
ID. *See* Intellectual disability (ID)
Illness, concept of, 70*f*
Image signal, 215
Imaging medicinal recognition, 157–158
Imitation task, 214
Immune system function, 67–68
Impairment, of speech utterances, 210
Independent component analysis (ICA), 62
Independent variables, 83

Index **269**

In-depth analysis, 7
Indian Patient Liver Dataset (ILPD), 187
Indian Satellite System (INSAT), 119–120
Indian Space Research Organization (ISRO), 119–120
Indirect emergency healthcare (EMH), 137
Infectious diseases, 69, 69*t*
Informational index, 198
Inherited diseases, 69, 69*t*
Initial cognitive development, 209
InnerEye, 19
Innovation, 158
Inpatient services, 138
Input excitation, 212
INSAT. *See* Indian Satellite System (INSAT)
Insulin, 71, 162
Integration of healthcare, 33
Intellectual disability (ID), 209–210
Intelligence system, 146
Intelligent algorithm, role of, 72
Intelligent digital wellbeing, 190
Intelligent health records, 158
Intelligent technology, 33
Internet of m-Health thing (m-IoT), 136–137
Internet of Things (IoT), 39, 181
  big data analytics, 115–116
  big data and, 156
    applications in healthcare, 158–159
  cardiotocography, 118
  health benefits of, 184–185
  in healthcare, 136–137
    application, 182–185
    cost leadership with quality of care, 148–149
    implementation, 140–149
    industry, 116–117
    and medical record, 142–143
    in patient delight, 148
    in supportive and utility services, 146–148
    and therapeutic facilities, 144–146
    value chain model, 140–149
  for medical coverage, 183
  software, 181*f*
  solutions' four phases, 184*f*
  techniques, 106–107
  in telemedicine system. *See* Telemedicine
Intrusive diarrheal carcinoma, 199
Invasive ductal carcinoma, 201
Inventory, retail store, 35
IoHTML, breast cancer in, 193–200
Isolated word recognition, 244
ISRO. *See* Indian Space Research Organization (ISRO)

J
Joint intelligence, 72

K
Keen camera, 154
K-fold cross-validation, 131
Kidney infection, 160
K-nearest neighbors (KNN), 100–101, 101*f*, 186, 193
KNeighborsClassifier, 100
KNN. *See* K-nearest neighbors (KNN)

L
Labeled data, 9
Larynx, 211
LDA. *See* Linear discriminant analysis (LDA)
Lead-I, 48
Lead-II, 48
Lead-III, 48
Learning, 74
Least mean square (LMS) algorithm, 54, 55*f*
Lifecycle for Programming Improvement, 193–194
Limb leads, 48
  placement, 49*f*
Linear discriminant analysis (LDA), 246
Linear predicative coding (LPC), 226–231, 246
  autocorrelation method, 230–231
Linear predictive cepstral coefficients (LPCC), 231–233, 233*f*, 246
Linear regression, 129*f*
Linear regression algorithm, 78, 79*f*
Linear time-invariant system, 231
Linen and laundry services, 147
Liver disease prediction, 187, 187*f*
LLMS algorithm. *See* Log least mean square (LLMS) algorithm
LMS algorithm. *See* Least mean square (LMS) algorithm
Lobular in situ (LCIS) carcinoma, 199
Location, for retail store, 34
Logical regression algorithms, 78
Log least mean square (LLMS) algorithm, 57–58
Lombard effect, 224
Long-term average power spectrum (LTAPS), 241–242
LPC. *See* Linear predicative coding (LPC)
LPCC. *See* Linear predictive cepstral coefficients (LPCC)
LTAPS. *See* Long-term average power spectrum (LTAPS)
LYmph Node Assistant (LYNA), 19
LYNA. *See* LYmph Node Assistant (LYNA)

M

Machine intelligence, in healthcare, 157–158
Machine learning (ML), 1, 2*f*, 19, 23, 36, 89–90, 117*f*, 157–158, 179–181, 180*f*
  algorithms, 1–2, 9, 24–25, 89–90
    fetal well-being using, 125–126
    in medicine, 77–82
    statistical algorithms and, 62
  architecture, 4–12
    data acquisition, 5
    data modeling, 9–12
    data processing, 5–9, 10*f*
    deployment, 12, 14*f*
    execution (model evaluation), 12, 13*f*
    phases of, 5*f*
  biomedical signal processing. *See* biomedical signal processing
  classification, 2–4
    reinforcement learning, 4
    supervised learning, 2–3
    unsupervised learning, 3–4
  and data mining techniques, 187
  diagnosis of disease. *See* Disease diagnosis system, machine-learning
  framework, 12–17
    Amazon machine learning, 15–16
    Apache mahout, 16
    features, 12–15
    Microsoft cognitive toolkit, 16–17
    Scikit-Learn, 16
    TenserFlow, 15
    types, 15–17
  in healthcare. *See* Healthcare, in machine learning
  integration, 91
  learning process in, 1
  mathematical-based definition of, 1
  methods, 179
  model, 7–9
  need for, 117–118
  radiology, 190–191, 191*f*
  in telemedicine/healthcare, 120–122
  types, 180*f*
Machine speech production system, 213, 213*f*
Machine speech recognition, 236
Machine vision system, 154
  application procedure, 154
  medicinal imaging, 154–156
Magnetoencephalography (MEG), 104
Mammogram, 90
Manufacturing, 36
MapReduce, 254–255
Massachusetts Institute of Technology—Beth Israel Hospital (MIT-BIH), 50, 54, 62

Mathematical model, 1–2
MATLAB, 93
Mean and variance normalization (MVN), 224
Mean square error (MSE), 53
Medical data, 31
Medical field, 113
Medical images, 19
Medical record keeping, 139
Medical record system, 142–143
Medicinal imaging, 154–156
  in digital health, 155–156
Medicinal services frameworks, 170
Medicine/drug, 70–73
  artificial intelligence in, 73–75
  different types of, 71
  discovering new, 71–72, 72*f*
  drug discovery process, 75–77
  machine learning algorithms in, 77–82
    artificial neural networks, 83
    convolutional neural network, 80–82, 80*f*
    expert systems, 82–83
    linear regression algorithm, 78, 79*f*
    logical regression, 78
    support vector machine, 78–80, 80*f*
  role of intelligent algorithm, 72
  routes for, 70, 71*f*
  working, 70–71
MedNet, 120
MEG. *See* Magnetoencephalography (MEG)
Mel-filter banks, 238–239
Mel frequency cepstral coefficient (MFCC), 226–228, 234–239, 240*f*
  delta coefficients, 239
  delta–delta coefficients, 240
  discrete cosine transform, 239
  fast Fourier transform, 237–238
  feature extraction, 235, 235*f*
  framing, 236–237
  Mel-filter banks, 238–239
  power spectrum density, 240–242
  preemphasis, 235–236
  windowing, 237
MEMS. *See* Microelectromechanical sensors (MEMS)
Mental diseases, 69, 69*t*
MFCC. *See* Mel frequency cepstral coefficient (MFCC)
Microchips, 32
Microelectromechanical sensors (MEMS), 144–145
Microsoft cognitive toolkit, 16–17
Minkowski metric, 100–101
m-IoT. *See* Internet of m-Health thing (m-IoT)

MIT-BIH. *See* Massachusetts Institute of Technology—Beth Israel Hospital
Mixed dataset, 29
MNet, 104
Model building, 118
MONALEESA-3, 202
MongoDB, 254—255
Motivation, 210—211
Moving average model (MA), 230
MSE. *See* Mean square error (MSE)
Multiple sensors, 145—146
MVN. *See* Mean and variance normalization (MVN)

N

Naive Bayes (NB)
    algorithm, 103—104
    classifiers, 204—205
Nasal cavity, 211
National Institute for Aging, 164
Natural defense system, 68
Natural intelligence, 72
NB. *See* Naive Bayes (NB)
Neural networks, 38—39, 245
Neurodevelopmental disorders, 209—211
Neurological disease, 91
    algorithms used in, 104*t*
    diagnosis, 104
Neuropathy, 160
NLMS algorithm. *See* Normalized least mean square (NLMS) algorithm
Noise reduction, 215
    technique, 223—225
        endpoint detection, 224—225
        frame blocking, 225—226
        spectral subtraction method, 224
        windowing, 226
Normal ECG signal, 50, 51*f*
Normalization of data, 8
Normalized least mean square (NLMS) algorithm, 54, 56*f*
Normative, definitions of, 67—68
Normativist, 67—68
Nosocomial infections, 146
Nutshell, in machine learning, 24—25

O

Obesity and overweight, 161
Obtrusive carcinoma, 200
Odor controlling, 147
1D-convolutional ANN model, 60
Online-based healthcare apps, 140
Open source projects, 253

Operation theater, 146
Option price, 36
Originality, in presented work, 170—171
Oscillatory nature, of speech waveform, 217
Outpatient services, hassle on, 138

P

Palbociclib plus letrozole, 203
Particle swarm optimization (PSO), 62
Passage configuration, 195
Patient booking system, 148
Patient-centric management system, 148
Patient medical record, 142—143
Patient monitoring system, 182*f*
Patient satisfaction, 138—140
PCA. *See* Principle component analysis (PCA)
PCM. *See* Pulse-code modulation (PCM)
PC vision, 157—158
Performance metrics, 97, 106*t*
Personal care, 158
Personal Information (PI), 164—165
PET. *See* Positron emission tomography (PET)
Pharmaceutical drug, 70
Pharmacotherapy, 70
Pharyngeal cavity, 211
Physical diseases, 69, 69*t*
Physiobank database, 60
Picture recognition task, 214
Pitch, 227
PLI noise. *See* Powerline interference (PLI) noise
PNN. *See* Probabilistic neural network (PNN)
Pooling layer, 82
Positron emission tomography (PET), 155
Powerline interference (PLI) noise, 52
Power spectrum density (PSD), 229, 240—242
Precision of algorithm, 98
Precordial leads, 50, 50*f*
Prediction
    accuracy, 9
    breast cancer diagnosis and, 105—106
    cancer, 196
    diabetes, 186, 186*f*
    disease diagnosis model for, 95—97
        classification technique, 96—97
        data preprocessing, 95
        performance metrics, 97, 106*t*
        training and testing data set, 95—96
    liver disease, 187
    machine-learning tools for, 91—92
Predictive analysis, 78
Preprocessing stage, 6
Primary activities, 140
Principle component analysis (PCA), 62

272 Index

Prior probability, 103
Private corporate industry, 140
Probabilistic neural network (PNN), 62
Problem-oriented medical records, 142
Professional speech pathologist, 210
Prognosis, 23–24, 31–33
Proper training, 215
Proposed algorithm, 58
Proposed method, 60–61
PSO. *See* Particle swarm optimization (PSO)
Public healthcare system, 38–39
Pulse-code modulation (PCM), 214–215
P waves, 61–62
Python, 92, 92*f*
  decision tree module in, 102

Q

Q-learning, 94
QRS complex
  detection technique, 61–62
  wave, 61–62
QRS duration, 50
Qualitative speech features, 227
Quantization, 216
Queue management, 138

R

Radial basis function neural network (RBFNN), 62
Radiography techniques, 154–155
Radiology machine learning, 190–191, 191*f*
Random forests, 246
  algorithm, 128
Raven surgical robot, 187, 188*f*
RBFNN. *See* Radial basis function neural network (RBFNN)
Recall metric, 98
Recording procedure, 214–215
Rectified Linear Unit (ReLU Layer), 81, 81*f*
Recurrent neural network (RNN), 60, 94–95
Recursion relation, 232–233
Regression, 3, 93–94
Reinforcement learning, 4, 30
  algorithms, 58, 94
  technique, 27–28, 28*f*
Reinforcement teaching, 180–181
Research dataset, 214*t*
Research methodology, 164–165
  trial setup, 164–165
Retail, 34
Risk prediction accuracy, 31–32
RNN. Recurrent Neural Network (RNN);.
  See Recurrent neural network (RNN)
Robots, surgery on, 187

Routes for medicine, 70, 71*f*
RR intervals, 61–62
Rule-based expert system, 32–33, 73

S

Sales, retail store, 35
Sampling, 216
Scaling of data, 8
Scikit-Learn, 16
SDLC. *See* Software development life cycle (SDLC)
SDLMS algorithm. *See* Sign data least mean square (SDLMS) algorithm
Secondary activities, 140
Security systems, 123–124, 255
Self-driving transport vehicles, 36–37
Self-inflicted diseases, 69, 69*t*
Semantic medical access (SMA), 137
Semimonitored education, 180
Semisupervised machine learning algorithms, 29, 94, 180
Sensitive Personal Information (SPI), 164–165
Sensitivity, 98
Sensors, 31, 122
Sentiment analysis, 37
Short-time Fourier analysis, 222
Sickness, concept of, 70*f*
Signal discontinuity, 226
Signal-to-noise ratio, 58
Sign data least mean square (SDLMS) algorithm, 55–56, 57*f*
Sine wave signal, 218
Sinusoidal signal, 217*f*, 218
  quantization process of, 218*f*
SIRTAR, 214
SMA. *See* Semantic medical access (SMA)
Smart dust, 144–145
Smart monitoring health devices, 32
Smart routing protocols, 35
Smart systems, 23, 33
Smoking, 164
Social communication, 209
Social diseases, 69, 69*t*
Social media, 37
Software development life cycle (SDLC), 193–194, 194*f*
Sophisticated algorithms, 98
Spectral deconvolution technique, 231
Spectral speech features, 226–227, 247
Spectral subtraction method, 223–224
Spectrograms, 219, 221
Speech classification, 244–246
  artificial neural network, 245

linear discriminant analysis, 246
random forests, 246
support vector machine, 245
Speech disfluency in communication, 209
Speech parameterization, 228—242, 234*t*
feature extraction, 228—229
linear predicative coding, 229—231
linear predictive cepstral coefficients, 231—233, 233*f*
mel-frequency cepstral coefficients, 234—239
weighted linear predictive cepstral coefficients, 233—234
Speech pathologies, 215, 235, 244
Speech pattern recognition system, 212
performance, 223—224
preprocessing stage, 243
types of, 244
Speech production, 211—213
components in, 211
human system, 212, 212*f*, 213*f*
by machine, 213, 213*f*
Speech recognition systems, 242—244. *See also* Automatic speech recognition system
Speech sample, 225—226
Speech signal processing, 215—226
cepstral analysis, 222—223
frequency analysis, 219—221
noise reduction technique, 223—225
sampling and quantization, 215—226
short time analysis, 221—222
short-time Fourier analysis, 222
time and frequency domain, signal in, 217—219
Speech signals, 221
Speech utterances, 209
Speed, 28
SPI. *See* Sensitive Personal Information (SPI)
Standard ECG signal, 48
Standard limb leads, 48
Statistical algorithms, 62
Statistical measures, 58—59
Statistics and Machine Learning Toolbox, 93
Stress, 172—173
Strong artificial intelligence system, 74
Supervised machine learning algorithms, 2—3, 25, 27*f*, 29, 58, 93—95, 98, 106—107, 179, 188, 190
Supportive and utility services, 139, 146—148
Support vector machine (SVM), 38—39, 78—80, 80*f*, 98—100, 99*f*, 107, 190, 245
Sustained phonation task, 214
Suturing automation, 187
SVM. *See* Support vector machine (SVM)
Systematic transportation system, 149
System modeling, 243—244

T

Tableau programming, 165
Telecommunication systems, 91
Telemedicine. *See also* Healthcare
challenges handling big data in, 123—125
fetal well-being using, 125—126
IOT network arrangement in, 123*f*
revolutionary effect of, 119—120
role of machine learning in, 120—122
Tele-monitoring, 119
TenserFlow, 15
Tertiary healthcare service providers, 138
Test data, 12
Therapeutic imaging, 155
Threatening tumor, 196
3D printing, 39
Thrombosis, 161—162
Thyroid disease, 38
Time-invariant filters, 52
Tiny segmented fragment, 225
Tomography/computed tomography, 201
Total mean square error, 230
Traditional diagnosis system, 90—91
Traditional programming, 25
Training dataset, 1, 6, 9, 29, 244
Transparency, 139—140
Transparent decision-making, 39—40, 40*f*
Transportation, 36—37
TTH-CAD distribution, 166—169, 166*f*, 167*f*, 171—172
analysis, 166—170
originality, in presented work, 170—171
recommendations and considerations, 171—172
Tumor phyllodes, 200
T waves, 61—62
Type 1 diabetes (T1D), 159—160, 163
Type 2 diabetes (T2D), 159—160, 162
Type-2 FCM clustering algorithm, 60—61
Type I error, 97
Type II error, 97
Typically developed (TD) child, 209—211

U

Unique identifiers (UID), 136
Unlabeled data, 6, 9
Unsupervised learning algorithms, 3—4, 6, 25, 27*f*, 30, 58, 94—95, 106—107

V

Vaccinations, 70
Value-based competition, 141—142

274 Index

Value chain model, 140–149
Velocity, 115
Veracity, 115, 251–252
Verbal communication, 209
Visualization, 118
Vocal cords, 211–213, 217
Voice identifiers, 244

W

Wavelet transform, 61–62
WDA. *See* Wearable device access (WDA)
Weak artificial intelligence system, 74

Wearable device access (WDA), 31, 137, 147, 253–254
Weighted linear predictive cepstral coefficients (WLPCC), 233–234
White Gaussian noise, 52
Windowing, 226, 237
WLPCC. *See* Weighted linear predictive cepstral coefficients (WLPCC)

Z

Zero reference potential, 48
Zero-sum competitions, 141

Printed in the United States
by Baker & Taylor Publisher Services